THE DISSIPATION OF ELECTROMAGNETIC WAVES IN PLASMAS

DISSIPATSIYA ELEKTROMAGNITNYKH VOLN V PLAZME

ДИССИПАЦИЯ ЭЛЕКТРОМАГНИТНЫХ ВОЛН В ПЛАЗМЕ

The Lebedev Physics Institute Series

Editors: Academicians D. V. Skobel'tsyn and N. G. Basov

P. N. Lebedev Physics Institute, Academy of Sciences of the USSR

Recent Volumes in this Series

Volume 57	Theory of Interaction of Elementary Particles at High Energies
Volume 58	Investigations in Nonlinear Optics and Hyperacoustics
Volume 59	Luminescence and Nonlinear Optics
Volume 60	Spectroscopy of Laser Crystals with Ionic Structure
Volume 61	Theory of Plasmas
Volume 62	Methods in Stellar Atmosphere and Interplanetary Plasma Research
Volume 63	Nuclear Reactions and Interaction of Neutrons and Matter
Volume 64	Primary Cosmic Radiation
Volume 65	Stellarators
Volume 66	Theory of Collective Particle Acceleration and Relativistic Electron Beam Emission
Volume 67	Physical Investigations in Strong Magnetic Fields
Volume 68	Radiative Recombination in Semiconducting Crystals
Volume 69	Nuclear Reactions and Charged-Particle Accelerators
Volume 70	Group-Theoretical Methods in Physics
Volume 71	Photonuclear and Photomesic Processes
Volume 72	Physical Acoustics and Optics: Molecular Scattering of Light; Propagation of Hypersound; Metal Optics
Volume 73	Microwave – Plasma Interactions
Volume 74	Neutral Current Sheets in Plasmas
Volume 75	Optical Properties of Semiconductors
Volume 76	Lasers and Their Applications
Volume 77	Radio, Submillimeter, and X-Ray Telescopes
Volume 78	Research in Molecular Laser Plasmas
Volume 79	Luminescence Centers in Crystals
Volume 80	Synchrotron Radiation
Volume 81	Pulse Gas-Discharge Atomic and Molecular Lasers
Volume 82	Electronic Characteristics and Electron – Phonon Interaction in Superconducting Metals and Alloys
Volume 83	Theoretical Problems in the Spectroscopy and Gas Dynamics of Lasers
Volume 84	Temporal Characteristics of Laser Pulses and Interaction of Laser Radiation with Matter
Volume 85	High-Power Lasers and Laser Plasmas
Volume 86	Superconductivity
Volume 87	Coherent Cooperative Phenomena
Volume 88	Cosmic Rays in the Stratosphere and in Near Space
Volume 89	Electrical and Optical Properties of III–IV Semiconductors
Volume 90	The Kinetics of Simple Models in the Theory of Oscillations
Volume 91	Lasers and Their Applications in Physical Research
Volume 92	The Dissipation of Electromagnetic Waves in Plasmas
Volume 93	Techniques and Methods of Radio-Astronomic Reception
Volume 94	Pulsed Neutron Research
Volume 95	Strong and Electromagnetic Interactions of Elementary Particles and Nuclei
Volume 96	Problems in the General Theory of Relativity and Theory of Group Representations
Volume 97	Exciton and Domain Luminescence of Semiconductors
Volume 98	Materials and Apparatus in Quantum Radiophysics
Volume 100	Research on the Condensation of Excitons in Germanium by the Microwave Method

In preparation:

Volume 99	Stimulated Raman Scattering

Proceedings (Trudy) of the P. N. Lebedev Physics Institute

Volume 92

The Dissipation of Electromagnetic Waves in Plasmas

Edited by

N. G. Basov

P. N. Lebedev Physics Institute
Academy of Sciences of the USSR
Moscow, USSR

Translated from Russian by

Donald H. McNeill

CONSULTANTS BUREAU
NEW YORK AND LONDON

Library of Congress Cataloging in Publication Data

Dissipaťsiîa elektromagnitnykh voln v plazme. English.
The Dissipation of electromagnetic waves in plasmas.

(Proceedings (Trudy) of the P. N. Lebedev Physics Institute; v. 92)
Translation of: Dissipaťsiîa elektromagnitnykh voln v plazme.
Bibliography: p.
1. Electromagnetic waves—Transmission—Addresses, essays, lectures. 2. Plasma (Ionized gases)—Addresses, essays, lectures. 3. Nonlinear theories—Addresses, essays, lectures. I. Basov, N. G. (Nikolai Gennadievich), 1922- . II. McNeill, Donald H. III. Title. IV. Series.
QC1.A4114 vol. 92 [QC665.T7] 530s [530.1'41] 82-5093
ISBN 0-306-10969-7 AACR2

The original Russian text was published by Nauka Press in Moscow in 1977 for the Academy of Sciences of the USSR as Volume 92 of the Proceedings of the P. N. Lebedev Physics Institute. This Translation is published under an agreement with VAAP, the Copyright Agency of the USSR.

Printed in the United States of America

PREFACE

This anthology includes articles on experimental studies of the interaction of high-power electromagnetic waves with collisionless plasmas and with electrons. The nonlinear interaction of waves with plasmas has been investigated both under free space conditions and in waveguides. A study of secondary-emission discharges was made in order to ascertain their possible effect on measurements in waveguides.

The results presented here on the interaction of high-power waves with plasmas and electrons are of interest to a wide range of physicists and engineers concerned with various questions on the interaction of electromagnetic radiation with plasmas, including microwave heating of plasmas and laser fusion.

CONTENTS

AN EXPERIMENTAL INVESTIGATION OF NONLINEAR DISSIPATION OF ELECTROMAGNETIC WAVES IN INHOMOGENEOUS COLLISIONLESS PLASMAS

G. M. Batanov and V. A. Silin

The methods of transferring energy from an electromagnetic wave to the electrons of a plasma by exciting various types of parametric instabilities are examined. Data are presented from an experimental study of the reflection and penetration of microwaves at an inhomogeneous collisionless plasma layer. Measurements of the energy of fast electrons, a determination of the combination frequencies, and modulation of the plasma density suggest that a $t \rightarrow l + s$ instability arises. It is established that the energy density of the plasma is not constant in time. This is manifested in the form of an almost periodic modulation in the fast-electron current, which exists throughout the time a microwave field is acting on the layer.

INTRODUCTION

The interaction of intense electromagnetic waves with inhomogeneous layers of dense plasma has attracted great interest in recent years. This interest is a result both of research on heating matter to thermonuclear temperatures with powerful laser pulses and of research on the propagation of intense radio waves in the Earth's ionosphere. In both cases we are dealing with a "dense" plasma in the sense that the frequency of the radiation ω_0 is less than the plasma frequency $\omega_{Le} = (4\pi ne^2/m)^{1/2}$, and with a collisionless plasma in the sense that the frequencies of pairwise collisions between electrons and ions ν_{ei} or between electrons and neutrals ν_{en} are small compared to the wave frequency. Recent research has shown that the interaction of a sufficiently intense electromagnetic wave with plasmas having these parameters causes an entire range of nonlinear processes which characterize both the dissipation of the wave energy in the plasma and the propagation of the wave (see [1-4]). Naturally, the dissipation and propagation processes are interrelated.

This article is devoted to a presentation of experimental data on the nonlinear penetration of microwaves through plasma layers whose dimensions are less than the mean free path for electrons and on the transformation of the incident wave energy into energy of the electron gas.

The authors' interest in the experimental study of the nonlinear transparency of layers was stimulated by the work of G. A. Askar'yan, V. I. Talanov, A. V. Gurevich, and V. P. Silin [5-8] which showed that plasma layers do not necessarily shield the incident wave field if it is large enough, that is, if the pressure $E_0^2/8\pi$ of the field approaches the gas kinetic pressure $n\varkappa(T_e + T_i)$ of the plasma. The experimental study of this problem began at the Lebedev Institute (FIAN) and the Institute of Radio Physics at Gorky (NIRFI) in 1968. The first reliable experimental data on the nonlinear nature of the wave penetration were published in 1971 [9, 10]. It was determined that the field penetrates plasma layers with maximum densities of 3-5 times the critical density over times comparable to the risetime of the incident pulse.

This effect was observed when the mean free path of the electrons was less than the linear dimensions of a layer [9] as well as in the opposite case [10]. Here the pressure due to the field of the incident wave was less than the plasma pressure. It was found that, at later stages of the interaction between the radiation and the layer, the density on the axis of the layer decreases and the field is enhanced in that region.

Later studies have shown (see, for example, [11, 12]) that the transparency of the plasma in the field of an intense wave is accompanied by other nonlinear effects. Thus, in the spectrum of scattered and transmitted radiation which goes through the layer it was possible to observe red satellites whose intensity increased nonlinearly with the intensity of the pump wave. Radiation was observed near the second harmonic of the pump wave. Langmuir and multigrid probe measurements of the characteristics of the electron gas showed that the average energy of the plasma electrons increases and that non-Maxwellian "tails" of fast electrons develop. Intense oscillations of the electron current in time were also recorded for this part of the distribution function [13].

The dissipation of electromagnetic waves in collisionless plasmas may be a result of the excitation of Langmuir potential waves in the plasma by one means or another. It is therefore natural to attempt to explain the entire set of experimental observations in terms of Langmuir turbulence of the plasma in an intense electromagnetic field.

CHAPTER I

TRANSFORMATION AND SELF-INTERACTION OF WAVES IN AN ISOTROPIC COLLISIONLESS PLASMA

The energy of fast electromagnetic waves can be dissipated in a collisionless plasma only as a result of their exciting slow electrostatic charge density waves which are then damped, for example, by Landau damping as their phase velocities approach the thermal velocities of the electrons or ions. In this regard, we can name two causes for transformation of electromagnetic waves into (potential) plasma waves: inhomogeneity of the plasma and nonlinearity of the wave interaction. In an inhomogeneous plasma the component of the electric field of the electromagnetic wave that is directed along the density gradient causes charge separation which in turn excites Langmuir waves near the plasma resonance ($\varepsilon = 1 - n/n_e = 0$). This process for excitation of plasma waves is called linear transformation.

With sufficiently high field strengths in the electromagnetic wave it is possible for density perturbations which grow in time to develop, along with high-frequency fields which are coupled to them. These wave excitation processes, which are caused by the forces resulting from the high-frequency pressure in inhomogeneous high-frequency fields, are referred to as nonlinear wave transformation.

The change in the plasma density caused by the high-frequency pressure is also the reason for the change in propagation conditions for waves in the plasma, that is, for their self-interaction.

We may speak of self-focusing or self-channeling of waves if changes in the density develop perpendicular to the direction of propagation of the wave, and of solitary waves or nonlinear transparency of a dense plasma ($n > n_c$) if the changes in density develop along the direction of propagation.

In general, besides striction, in a collisionless plasma a thermal mechanism for the self-interaction is possible. This may occur when the energy release from linear or nonlinear transformation is nonuniform within the plasma layer because of a nonuniform field in the incident wave. If the electron component of the gas is heated, then the region in which heating occurs and, therefore, in which the pressure increases, may be localized because of the imbalance between the fluxes of electrons from the hot region and from the cold regions, that is, because of conservation of quasineutrality in the plasma. As a result, the so-called substitution wave regime develops [14, 15] and the plasma density falls in the heating region.

In the following discussion we shall limit ourselves to a brief exposition of the basic theoretical results on transformation and self-interaction of waves in plasmas.

1. Linear Wave Transformation and Effects Related to the Excitation of Plasma Waves

The first theoretical investigations of linear wave transformation were made in connection with research on the propagation of radio waves in the Earth's ionosphere and on the radio emission from the sun and planets [16, 17]. Later on, the mechanism of linear transformation was studied in a number of papers (see, for example, the reviews [18, 19]).

In an inhomogeneous plasma if the angle of incidence of the wave is nonzero, then the trajectory of a transverse wave will be bent and at the reflection point ($\varepsilon = \sin^2 \vartheta$) there will be a substantial component of $\mathbf{E}$ parallel to ∇n. Because of this, energy is transferred by subbarrier penetration to the plasma resonance point, where the growing electric field already has the form of Langmuir oscillations or, more precisely, plasma waves, if the temperature of the plasma is taken into account. In the excitation region their wave vector is given by $k_l \simeq [(\varkappa T_e/m\omega^2)L]^{-1/3}$.

The transformation efficiency, defined as the ratio of the energy flux going into plasma waves to the incident energy flux, depends on the size L of the plasma inhomogeneity and on the angle of incidence ϑ. From physical considerations it is clear that for strictly normal incidence no energy is converted to plasma waves since there is no E_z component of the field (we have the band near ω_0 in mind) while for grazing incidence the transformation efficiency is also small since the reflection point is far removed from the point $\varepsilon = 0$. As

Piliya [20] has shown, for $L \gg \lambda$ the transformation coefficient may reach 0.4 for an optimum angle of incidence.

Electron plasma waves may propagate in a rarefield plasma in accordance with the dispersion relation $\omega^2 = \omega_{Le}^2 + 3k^2 v_{Te}^2$ and may be damped by collisionless mechanisms. In the linear case (for infinitesimal wave amplitudes) this effect is known as Landau damping and obeys the following formula for the damping coefficient:

$$\tilde{\gamma}_0 = \sqrt{\frac{\pi}{8}} \frac{\omega_{Le}^4}{k^3 v_{Te}^3} \exp\left(-\frac{\omega_{Le}^2}{2k^2 v_{Te}^2} - \frac{3}{2}\right).$$

When the amplitude of the field is increased, the linearity fails. In collisionless damping this shows up as a reduction in or suppression of damping. Depending on the potential Φ in the plasma wave, the damping varies as (see [21])

$$\tilde{\gamma} = \frac{\tilde{\gamma}_0}{1 + (e\Phi)^{3/2}/(\sqrt{m}\,\varkappa T_e \nu_{eff} \lambda)}.$$

This reduction in the damping is a result of the formation of a "plateau" in the electron distribution function because of the acceleration of particles moving synchronously with the waves.

Including the finite amplitude of a field in the resonant region in an inhomogeneous plasma also leads to nonlinearities. In 1964 Gil'denburg [22] showed that the distribution of the Langmuir field and of the electron density (neglecting spatial dispersion) experiences a discontinuity and becomes nonunique. Either there is jump in the density or a layer with zero permittivity is formed. In the case of a jump, in particular, hysteresis effects may occur; that is, depending on the direction of the preceding change in the field, the resonance point is either on a steep or on a flat part of the density change. Then there is a reduction in the magnitude of the resonant field strength.

The problem of the field distribution and density in a resonant region including a weak spatial dispersion has recently been solved numerically [23]. It was shown that the solution in this case is discontinuous. The solution oscillates through zero to the left of the transition point ε while to the right of this point the field and density perturbation fall monotonically. Estimates of the transformation coefficient for this case indicate that it decreases as the field is increased.

The effect of nonlinear processes on the amplification of the field is also important when there is no stationary state for the density distribution. A limit in the field amplitude because of nonlinearity in the electron oscillations was found in [24] and it was shown that already in the nonstationary stage the amplitude of the field may be considerably less than if the ion motion is neglected.

2. Parametric Excitation of Plasma Waves and "Anomalous" Plasma Heating

The idea that nonlinear wave interactions could be the reason for intense dissipation of electromagnetic waves in a collisionless plasma was expressed as far back as 1944 [25]. Later on, nonlinear transformation was used to explain the radio emission from the sun and planets [26]. Various aspects of the problem of nonlinear wave interactions were formulated in a number of papers [27-31] and at present intensive theoretical research on this topic is under way (see, for example, the reviews [1, 32-35]).

In quantum terms the physical nature of the processes of nonlinear transformation can be specified as stimulated combination scattering of waves and stimulated scattering of waves by charged particles. The characteristic process may be said to be stimulated Mandelshtam–Brillouin (Brillouin) scattering (SMBS) when the pump wave is scattered on acoustic waves while the intensities of the acoustic waves and of the scattered electromagnetic waves are nonlinearly amplified compared to spontaneous processes. For a plasma, combination scattering processes can be described in terms of three-wave decay in which the initial wave with frequency ω_0 and wave vector $\mathbf{k}_0$ excites perturbations with frequencies ω_1 and ω_2 and wave vectors $\mathbf{k}_1$ and $\mathbf{k}_2$ in a way such that the conservation laws $\omega_0 = \omega_1 + \omega_2$ and $\mathbf{k}_0 = \mathbf{k}_1 + \mathbf{k}_2$ are satisfied. In the three-wave interaction scheme SMBS is described in the form of the decay of a photon (t) into a photon (t') and sound (s): $t \to t' + s$. Among the processes that are most important for nonlinear wave transformation in an isotropic plasma with a density close to critical, we also include the decay of a photon into a plasmon (l) and sound ($t \to l + s$) or into two plasmons ($t \to l + l'$), the decay of two photons into two plasmons ($2t \to l + l'$), and the stimulated scattering of photons on ions (i) with formation of a plasmon ($t \to l + i$).

TABLE 1

Instability	Decay type (quasidecay)	Range of existence	Threshold in inhomogeneous plasmas
Aperiodic	$2t \rightarrow l + l'$	$\omega_{Le} \gtrsim \omega_0$	$\frac{E_0}{E_p} > \left(\frac{4\nu_{ei}}{\omega_{Le}} + \frac{2}{kL}\right)^{1/2}$
Periodic (potential)	$t \rightarrow l + s$	$\omega_{Le} \lesssim \omega_0$	$\frac{E_0}{E_p} > \left(8\pi \frac{m}{M}\right)^{1/2} (kL)^{-1/2}$
Decay into two plasmons	$t \rightarrow l + l'$	$\omega_{Le} \simeq 1/2\omega_0$	$(v_E/c)^2 (kL)^{4/3} > 1$
SMBS	$t \rightarrow t' + s$	$\omega_0 \geqslant \omega_{Le} > \frac{M}{2m} \frac{v_{Ti}^2}{c^2} \omega_0$	$\frac{E_0}{E_p} > \frac{\omega_0}{\omega_{Le}} (kL)^{-1/2}$
SCS	$t \rightarrow t' + l$	$\omega_0 > 2\omega_{Le} > \frac{v_{Te}^2}{c^2} \omega_0$	$(v_E/c)^2 kL > 1$

In the classical theory these processes can be viewed as a parametric instability in a system of coupled oscillations (in which capacity the characteristic oscillations of the plasma serve), and they are coupled by the pump wave via the nonlinear properties of the plasma. In this case the physical picture of these phenomena may be interpreted in terms of an rf quasipotential, that is, in terms of the rf pressure averaged over the period of the rf field.

One specific feature of the parametric excitation of waves is the existence of threshold fields determined by the magnitude of the dissipation in the system [34, 36]. It should be noted that in a homogeneous plasma with a low collision frequency the threshold fields generally correspond to the condition $E_0 \ll E_p$ [$E_p = (3\varkappa T_e m\omega_0^2 \times e^{-2})^{1/2}$ is the characteristic plasma field]. Inhomogeneities in the plasma layer or in the pump wave field lead to a rise in the threshold fields because of the limited size of the wave interaction region in which they satisfy the conditions for synchronism [37]. Without dwelling on the various forms of instability, we include a table with their threshold fields and domains of existence as determined from the results in [34, 35, 37-39] (see Table 1).

The dissipation of the energy of electromagnetic waves by means of the development of a parametric instability can be described with the aid of the usual expressions for the rf conductivity of plasmas if the pair-wise collision frequency is replaced by some effective collision frequency [34]. This quantity evidently characterizes the efficiency of the nonlinear wave transformation process. For pump wave amplitudes close to threshold it is proportional to the excess over the threshold field E_{0t}:

$$\nu_{ef} \sim (E_0^2 - E_{0t}^2).$$

At high field amplitudes the growth ν_{ef} slows down as has been shown in some "computer experiments" [40].

An important topic in a discussion of parametric excitation of oscillations is the question of the mechanism for limiting the amplitude of unstable oscillations and the mechanism for energy transfer from the oscillations to the charged particles in the plasma. Despite a large amount of work on this question it is still far from solved. Here we shall consider some typical processes which are presently being discussed in the literature.

In an isothermal plasma near the plasma resonance, only Langmuir waves can be excited. This occurs either because of conversion of electromagnetic waves on ions or because of the aperiodic (two-stream) instability. It appears that in this case the stimulated scattering of parametrically excited Langmuir waves on the ions determines the ultimate amplitude of the turbulent fluctuations in the plasma (see [41-43]). The Langmuir turbulence is nonstationary. We have a similar situation when a photon decays into two plasmons [44, 45]. In the case of a nonisothermal plasma the situation is more complicated. Thus, in [46, 47] it has been shown that the parametric instability can be stabilized as a result of a sequence of decays of Langmuir waves into ion-acoustic and Langmuir waves and that the so-called dynamic turbulence regime can be initiated.

In [48] stimulated scattering of ion-acoustic oscillations by ions was examined as a nonlinear mechanism for limiting the wave amplitude. It is interesting to note that various aspects of the interaction of Langmuir and ion-acoustic waves have been modeled with the aid of a nonlinear rf transmission line [49]. There the nonstationarity of Langmuir turbulence was convincingly demonstrated.

It is easy to see that the nonlinear pumping of Langmuir waves leads to the excitation of weakly damped long wavelength oscillations. In this regard the further evolution of the waves was one of the most vital problems in Langmuir turbulence. Studies of the stability of long wavelength Langmuir oscillations showed that a unique self-interaction process occurs that is analogous to self-focusing, in which there is a drop in the plasma density and amplification of the wave field in a localized region, the so-called collapse of Langmuir waves [50] caused by the development of the modulation instability [35, 51]. Collapse was invoked in [52] as a reason for dissipation of Langmuir waves to explain the field damping at a singularity during linear wave transformation. However, the question of rapid transfer of the energy of Langmuir waves to the plasma electrons during collapse is still open since it has not been determined whether the self-compression of the collapsing cavity is limited or not.

Including the inhomogeneity of the plasma evidently allows us to point out one mechanism for energy transfer from plasmons to the electron component of the plasma without relying on collapse. Actually, it has already been pointed out in [53] that Langmuir waves propagating in layers of rarefied plasma are strongly Landau damped because of a reduction in the phase velocity of the waves.

The heating of electrons in an inhomogeneous plasma in the case of a single mode of Langmuir waves has been examined in [54, 55] and, in the case of a broad spectrum of waves, in [56]. In the first case it was shown that a beam of electrons was formed and in the second, that a broad energy group of fast electrons was formed. It is known that substantially above threshold levels such parametric processes as the conversion of electromagnetic waves into Langmuir waves by ions and the decay of ion-acoustic and Langmuir waves have maximum growth rates for excitation of sufficiently short wavelength plasma waves ($kr_{De} \simeq 0.2$–0.3). If we neglect the change in the electron distribution function then it is evidently possible to suppose that a large part of the energy of these short wavelength oscillations will be transferred to the electrons by Landau damping and to a greater extent in the case of an inhomogeneous plasma layer.

3. Possibility of Nonlinear Transparency in Supercritical Plasmas

According to the linear theory of the interaction of radiation with plasmas, a supercritical plasma will screen out an incident transverse electromagnetic wave over a length $c/(\omega_{Le}^2 - \omega_0^2)^{1/2}$. However, a strong electromagnetic field can modify the electron density distribution in the plasma and thus change the conditions for penetration of radiation into the plasma.

If nonlinear transparency develops as a consequence of changes in the refractive index caused by a motion of the electron density perpendicular to the direction of propagation of the wave, then it is customary to speak of the possibility of self-focusing of the wave or a quasi-waveguide propagation [5, 6, 57]. This effect is easily interpreted in terms of ray optics with the aid of total internal reflection from the boundary of the quasi-waveguide channel. Here the condition for compensation of the intrinsic (diffractive) divergence of the ray has the form [57]

$$\frac{cE_0^2}{8\pi} \geqslant \frac{(1.22\lambda_0)^2}{S_0} \frac{c}{64 N_0 N_2},$$

where S_0 is the cross-sectional area of the beam, N_0 and N_2 are the linear and nonlinear parts of the refractive index, and $N = N_0 + N_2 E_0^2$. This condition is valid only for small divergence angles ($\theta = 1.22\lambda_0/(2RN_0) \ll 1$) and in fact determines the threshold value for the energy flux.

For a plasma the dependence of the dielectric constant of the plasma on the wave field is given by [58]

$$\varepsilon = 1 - \frac{\omega_{Le}^2}{\omega_0(\omega_0 + i\nu)} \exp\left(-\frac{E^2}{E_p^2}\right).$$

For thermal processes involving collisions of electrons with ions and atoms the magnitude of E_p is smaller by a factor of $(M/2m)^{1/2}$ than for striction processes.

Talanov and Litvak [6, 59], during an examination of two-dimensional traveling TE and TM waves in plasmas, found spatially localized solutions for the field of the form

$$E_z(x) = \frac{E_m}{\cosh[\varepsilon_\infty (h^2 - k_0^2) x]},$$

where k_0 is the wave number of the incident wave and ε_∞ is the value of the dielectric constant at infinity.

In a subcritical plasma, formation of a waveguide channel may occur at any wave power. Only the width of the channel in x and the shape of the field distribution within it depend on the power level. In a supercritical plasma ($\omega_{Le}^2/\omega_2^0 > 1$) there is a threshold for this effect given by

$$E_0^2 \geqslant E_p^2 \frac{2(\omega_{Le}^2/\omega_0^2 - 1)}{\omega_{Le}^2/\omega_0^2}.$$

Thus, the threshold field for a very dense plasma is $\sqrt{2}E_p$ and decreases, approaching zero, as the density approaches the critical value.

It should be noted that the above condition is the condition for existence of a waveguide channel in a supercritical plasma.

However, the field strength at which the wave pressure can "dig" such a channel in the plasma layer evidently may be of order E_p.

Quite recently an important result on self-focusing of waves in plasmas was obtained by Litvak and Fraiman [60] in which they were able to include the effect on self-focusing of field singularities near $\varepsilon = 0$ in the presence of a field component along the density gradient.

There is another class of phenomena which can cause nonlinear transparency of a plasma which involves modulation of the plasma density in the propagation direction of the incident wave. The possibility of stationary periodic layers of dense plasma and layers of electromagnetic field with a characteristic dimension roughly equal to half the wavelength of the radiation was first established in [61].

Later a theory of the nonlinear propagation of electromagnetic waves in plasmas was developed which described, in particular, the nonlinear transparency of a layer of supercritical plasma [7, 8, 62-67].

V. P. Silin, in an examination of the normal incidence of a wave on a conductor or plasma, showed [8] that as the field of the incident wave approaches the critical value [equal to $E_c = 2^{-3/2}(\omega_{Le}/ck_0)E_p$, where $k_0 = \omega_0/c$] the depth of the nonlinear skin layer may be much greater than the linear skin layer. In the region where the field is large, slowly damped (in space) nonlinear oscillations of the field develop. Their distribution with respect to z is described by nonharmonic spatial oscillations in the self-induced transparency region. For significant penetration into the conductor (or in a very dense plasma region) the penetration converts to an aperiodic exponential decay of the field over a distance of the same order of magnitude as the linear skin depth. This picture naturally characterizes the stationary state. Here the reflection point of the wave, as in the studies by other authors [63], moves deeper into the layer as the field E_0 is increased.

A numerical calculation [64] for homogeneous layers of supercritical plasma with a finite value of $\omega_{Le}/\omega_0 > 1$ shows that the dependence of the reflection coefficient of a layer on the thickness is quite different from the linear approximation. Specifically, in the nonlinear case the plasma is more transparent and the reflection coefficient R^2 varies periodically with the layer thickness as it approaches zero (the transmission coefficient varies in accordance with R since the problem is nondissipative). This transparency corresponds to the resonant transparency of plane layers in which the dielectric permittivity is an alternating function of the coordinate.

A more general case was recently calculated by Gil'denburg [65] who examined the nonlinear transmission of a wave through an inhomogeneous plasma layer. It was shown that the amplitude dependence of the transmission coefficient of the wave is not unique and, in addition, several intermediate "resonance" branches were obtained which correspond to transparency of a layer with finite thickness.

The reflection coefficient of a plasma layer taking the inhomogeneity of the layer into account was also computed in [66].

In the above references, only the steady-state distributions of the density and field were examined. It might be assumed that the time for this condition to be established is at least λ_0/v_s, where v_s is the speed of sound in the medium.

The dynamics of the interaction of an electromagnetic wave with a layer of dense plasma was examined in [67]. There the author analyzed the penetration pattern of the field into a layer of supercritical plasma and showed that ultimately an electroacoustic soliton propagates in the plasma (that is, a density perturbation with an electromagnetic wave trapped in it). In this case the time for the wave to penetrate through the layer is determined in order of magnitude by the sound speed.

More rapid penetration of a wave through a plasma layer can evidently occur when a parametric instability develops. It is known that the sound waves which arise during a parametric instability modulate the plasma density along the field vector of the incident wave. We have noted [10] the fact that when the modulation depth is sufficient in this case a regular or chaotic structure may appear which is analogous to a set of metal slabs oriented with their faces perpendicular to the electric field of the wave. Such a structure is transparent to the radiation and has a reflectivity of less than unity.

In [68] the transparency of a plasma layer was calculated under the assumption that the density modulation along the field vector of the incident wave is caused by rf pressure which is a result of the modulational instability of parametrically excited Langmuir waves. The authors found the characteristic period of the density modulation in the field of oppositely propagating Langmuir waves and showed that there is a threshold field for transparency,

$$E_m^2/E_p^2 \geqslant {}^1/_{64}\ (2M/m)^{1/2}\ |\ 1 - n/n_c\ |^{1/2}.$$

Here E_m is the maximum value of the field at the boundary of the plasma layer.

CHAPTER II

EXPERIMENTAL RESULTS FROM A STUDY OF NONLINEAR TRANSPARENCY

In experiments on the nonlinear transparency of plasmas and on the change in the trajectory of electromagnetic waves in them, it is assumed that the size of the plasma must be much greater than the wavelength of the radiation. In laboratory studies we can actually expect to produce a plasma with a maximum size on the order of 1 m with an electron density of 10^{12} cm^{-3}. Thus, a microwave oscillator with a wavelength $\lambda_0 \ll 1$ m is required, while, if it is planned to expel plasma from the region occupied by the wave by means of the pressure of a transverse wave, then it is necessary to produce an energy flux such that $E_0^2/8\pi \gtrsim n\varkappa T_e$. Taking a plasma produced by a spark source with a temperature $T_e \simeq 10$ eV as an example and keeping in mind that in free space conditions the minimum focal size of the field is $\lambda_0/2$, we find that the oscillator power capable of creating a field pressure equal to the plasma pressure at the critical point is

$$P_0 = \frac{cE_0^2}{8\pi}\,\pi\frac{\lambda_0^2}{4} = \frac{n_c\varkappa T_e}{4}\,\pi\lambda_0^2 c,$$

where $n_c = \pi mc^2/e^2\lambda_0^2$.

It is clear that under these conditions the oscillator power is independent of the wavelength and is given by

$$P_0 = \frac{\pi^2 mc^3 \varkappa T_e}{4e^2},$$

which yields 500 kW for $T_e = 10$ eV. Consequently, the most appropriate band is $\lambda \simeq 10$ cm. At this wavelength it is relatively easy to generate powers of several megawatts and, thus, to make the field pressure exceed the plasma pressure, even if the density is an order of magnitude above critical.

Based on these considerations a microwave generator using a magnetron for the 10 cm band was constructed which could yield infrequently repeating pulses of microwave radiation with powers of up to 2.5 MW and a controlled length of from 0.5 to 25 μsec. The angular frequency ω_0 of the oscillations was roughly 2×10^{10} sec^{-1}. A detailed description of this apparatus is given in [11]. Here we shall give only a schematic illustration of the experimental conditions (Fig. 1) and point out the basic parameters of the plasma layer.

The plasma layer was produced by means of four Bostick gun type spark injectors [69] each of which injects a plasma stream toward the center of the vacuum chamber. Diagnostic measurements showed that the plasma layer is formed according to the following pattern. At times sufficiently close to the startup of the guns, a plasma with a near-critical density is found near the injectors and moves toward the center of the vacuum vessel. Then the plasma bunches come together and a plasma layer is formed. Within this layer the critical density boundary moves at first toward the emitter (left in Fig. 1). After 10-15 μsec it begins to move in the opposite direction. By appropriate choice of the time the oscillator is turned on it is possible to achieve practically normal incidence of the wave on the plasma layer.

The thickness of the layer (to the level of the critical density) was about 30 cm, the characteristic size of the inhomogeneity was 15 cm, and the peak density was 3-5 times the critical density. The microwave radiation was delivered to the layer in the form of a slightly divergent flux whose perpendicular dimension was about 15 cm.

4. Measurement Technique

In these experiments the field of the incident, transmitted, and reflected waves were measured at various parts of the device. Detectors with crystal and vacuum tube rectifying elements were used for this purpose.

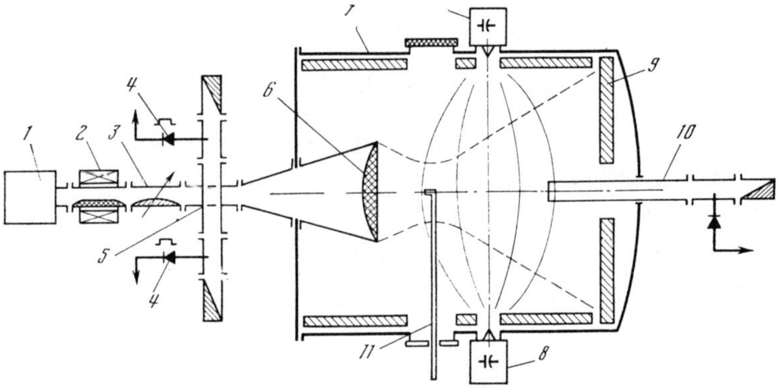

Fig. 1. A diagram of the experimental apparatus: (1) microwave oscillator; (2) ferrite rectifier; (3) absorbing attenuator; (4) pickups for the envelope of the microwave radiation; (5) directional coupler; (6) focusing lens; (7) vacuum chamber; (8) plasma injectors; (9) absorbing elements; (10) receiver waveguide; (11) multigrid probe.

The high-power microwave oscillator signals were measured with D-605 linear detectors or with 6S5D vacuum triodes with disk outputs. The low-power cold measurements were made with quadratic detectors, which made it possible to use instrument amplifiers, or with a set of series P5 instrument receivers.

Several types of probes were made which could be used either as electric microwave probes or as single Langmuir probes. These probes had a small electrical antenna oriented parallel to the $\mathbf{E}_0$ vector of the incident wave. The antenna output was connected to a coaxial cable. A four-antenna probe which was introduced from the end of the vacuum vessel was used to measure the distribution of the high-power microwave field in the plasma. The probe was made in the shape of the letter T out of stainless steel tubes. The antennas were located on one side of the cross member and the other side served as an electrical "counterweight" (Fig. 2). This probe could also be used as a single Langmuir probe for measuring the density. Such measurements also confirmed that the antennas were blocked off by dense plasma.

The electron energy was measured with the aid of a multigrid probe (Fig. 3) by the retarding potential method. The probe electrodes were mounted in a metal screen so that they could be inserted directly into the microwave-plasma interaction region. The electrode leads from the probe passed through a metal tube about 50 cm long mounted in the chamber perpendicular to the electric field of the incident microwave radiation. To ensure minimal distortion of the field and plasma in the measurement region it is desirable that the probe be as small as possible, at least compared to the wavelength of the incident radiation.

The probe was operated in two regimes: with a negative potential (variable) on the collector and ion cutoff at grid 2 or with a positive potential (+200 V) on the collector and electron cutoff at grids 2 and 3 (a variable negative potential). The second regime is characterized by large collector currents because the space-charge current limitation has less effect (the effective gap size is reduced).

5. Measurements of the Microwave Fields of the Transmitted and Reflected Waves

In these measurements it was possible to realize two interaction regimes between the microwaves and the plasma layer. In the first, a layer of overdense plasma was prepared before the microwaves were turned on and in the second, the microwaves were turned on 2-3 μsec before the plasma layer had formed. That the microwaves were not cut off in the second regime (at powers above 500 kW) and there was a slight increase in the transmitted signal at the center of the beam is not unexpected since under these conditions some nonlinear mechanism must certainly exist for dissipation of the waves and heating of the plasma.

In the first regime (i.e., normal incidence on a layer of overdense plasma) the amplitude of the incident wave is found to have a distinct effect on the penetration process. In general, because of thermal spreading of the layer, after a certain time the layer becomes transparent even to a weak signal. Increasing the power causes little change in the time at which the layer becomes transparent if the power does not exceed 7-10 kW. For more intense signals a gradual reduction in the time for the wave to penetrate is observed as its power is increased. This effect was observed with both a dowel antenna and a movable waveguide.

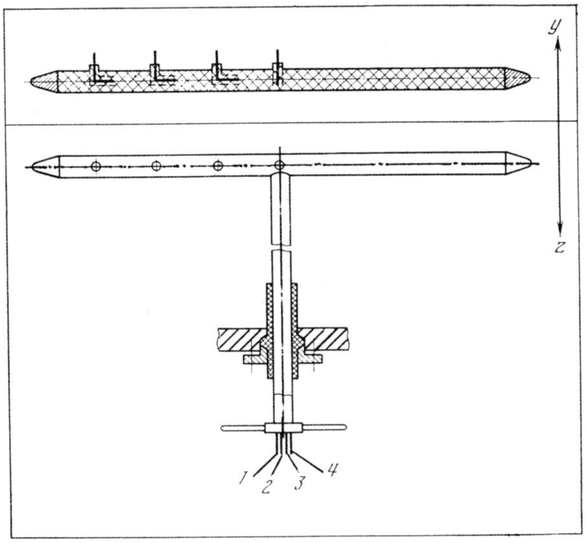

Fig. 2

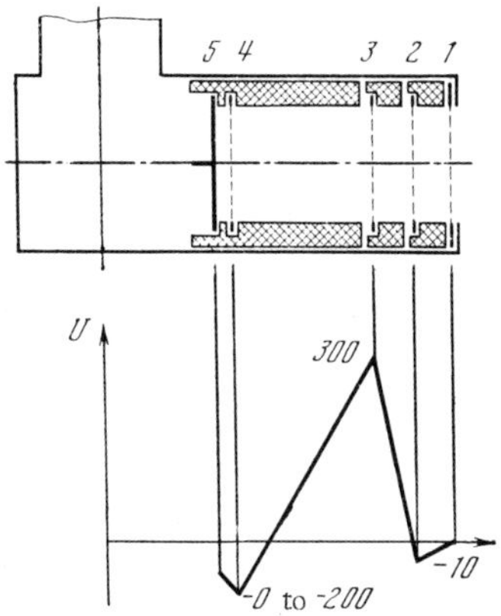

Fig. 3

Fig. 2. The design of the four-antenna microwave probe with a symmetrizing "counterweight." The numbers denote the numbers of the antennas.

Fig. 3. The design of the multigrid probe for analyzing the electron energy and the potential contours in it.

From the dependence of the characteristic time for transparency of the layer on the incident wave power (Fig. 4), it follows that the onset of the nonlinear process occurs at a level of roughly 10 kW or 30 W/cm. The transparency time is defined here as the time over which the amplitude of the transmitted signal reaches $(I - I/e)E_0$, where E_0 is the field in the absence of plasma. As the power of the pump wave is varied, both the transparency time of the plasma layer and the manner in which the wave penetrates the layer will change. If at a relatively low power level the wave is damped in the interior of the layer and this distribution gradually propagates from the boundary of the layer into its interior, then at powers above 300 kW the wave permeates the entire depth of the plasma (Fig. 5). In that case, at the beginning of the layer (in rarefied plasma) the field distribution is the same as at a lower power level, while in the dense plasma, not damping, but penetration of the signal is observed. This penetration is established very rapidly, over a time comparable to the risetime of the microwave pulse. It can be definitely stated that 0.2 μsec after the high-power pump wave is turned on, its field penetrates to densities of $n = (2\text{-}3)n_c$.

The following circumstance should be noted. On the oscilloscope traces and on graphs constructed from the readings from the dowel antenna the ordinate is the detector current which is proportional to the field in the cable from the antenna; however, that is not the field in the plasma. The plasma field can be determined if we take the coupling coefficient of the antenna with the radiation into account. This coefficient tends to zero as the plasma resonance is approached. Thus, the actual field in a resonance region of the plasma may be much greater than that shown in Fig. 5 (0.1 and 0.3 times the maximum value).

By varying the maximum plasma density in the layer it was established that denser layers of plasma become transparent somewhat later. This effect can be illustrated with the aid of oscilloscope traces of the signals passing through the plasma (Fig. 6) and received by a horn located at the rear wall of the vessel on its axis.

Besides the change in the wave penetration as the power is increased, there is a change in the character of the reflection of the wave from the layer. The reflection of the wave at 180°, that is, back into the antenna horn, was measured with the aid of a directional coupler. At low powers the magnitude of the reflected signal approaches the incident signal ($R = 1$), does not change for some time (5-8 μsec), and then decreases, possibly because of thermal expansion of the layer and changes in the form of the reflecting boundary. As the power is increased, the relative magnitude of the reflected signal falls and the time interval over which the reflection remains unchanged is shortened (Fig. 7). The reduction in the reflectivity as the power of the incident wave is increased is an indication that either dissipation or "fast" transparency is occurring, while the

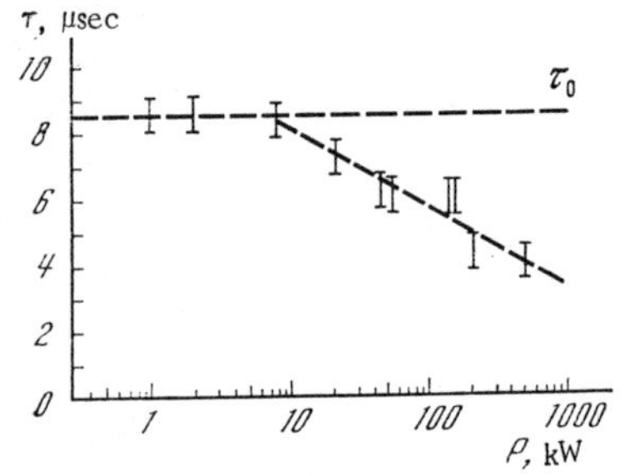

Fig. 4

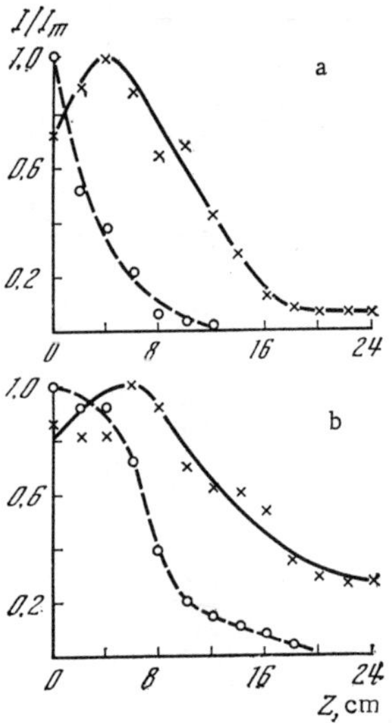

Fig. 5

Fig. 4. The dependence of the time for the layer to become transparent on the incident wave power.

Fig. 5. The distribution of signals recorded by the dowel antenna in the region between the emitter and the plane of the plasma guns: (a) at 0.2 μsec after the beginning of the microwave pulse; (b) 5 μsec after the start of the microwave pulse. The smooth curves correspond to a power of 500 kW, the dashed, to 2 kW.

shortening of the interval over which the reflectivity is constant implies that large changes are taking place in the form of the reflecting boundary of the plasma.

In addition to the changes in reflection, depending on the increase in field strength of the incident wave, intense fluctuations in the density of charged particles are observed (Fig. 8). Of particular interest is the fact that for 2-4 μsec after each density drop phase, the density rises again to at least its original value. Only after 2-4 μsec do the density oscillations die out and the average values become smaller than the unperturbed values.

In conclusion we note that both "fast" and "slow" changes in the transparency properties are observed at relatively low field strengths. In fact, both a reduction in the transparency time of the plasma layer and changes in the nature of wave reflection are detected at powers $\gtrsim$10 kW ($E_0/E_p = 0.03$). As the power is increased to 300-500 kW ($E_0/E_p \simeq 0.3$), $E_0^2/8\pi$ still remains substantially below $n\varkappa T_e$.

6. Distortions in the Scattered Radiation Signals

In these experiments we have studied the spectrum of the radiation which penetrates the plasma layer. The microwave signal received by the open end of a rectangular waveguide positioned at the center of the layer on the axis of the vacuum chamber was fed to a single pulse heterodyne spectrum analyzer (type S4-14). The spectrum is analyzed because the second intermediate frequency signal is fed into a delay line with frequency dispersion. Of course, this method can only be used to analyze rather short pulses since a frequency scale is transformed to a time scale during the measurements. The intermediate frequency amplifier was strobed by a 2-μsec-long pulse and the delays were chosen so that the spectrum of the portion of the microwave pulse that penetrated the layer between the second and fourth microseconds after the pulse began could be analyzed (Fig. 9).

The measurements were done as follows: the spectrum of a pulse incident on the receiver in the absence of plasma was measured and in the next pulse the plasma injectors were turned on with the same gain and coupling coefficient for the microwaves. It was found that the intensity of radiation at the fundamental frequency in the plasma decreases compared to the intensity of the microwave generator. In addition, extra lines appear (Fig. 9). The reduction in the amplitude of the fundamental (due to partial reflection of the microwave

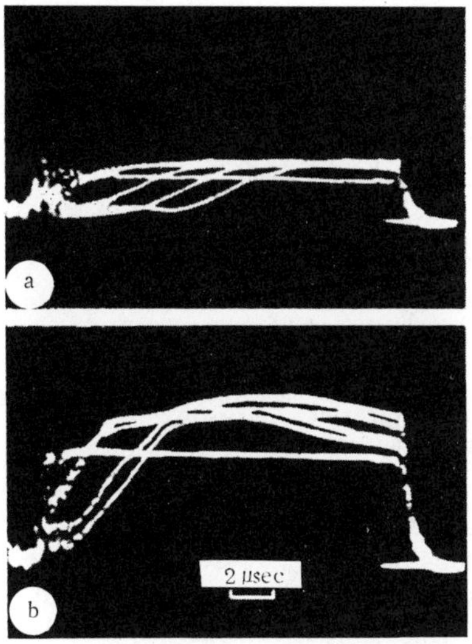

Fig. 6

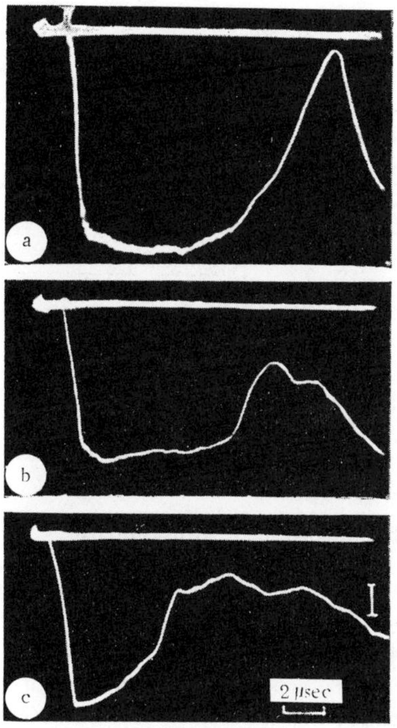

Fig. 7

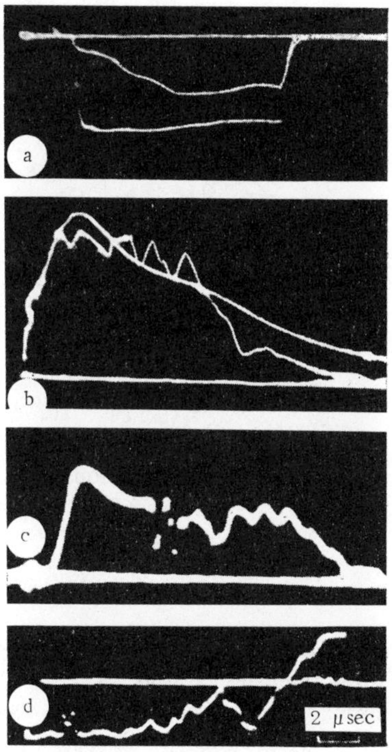

Fig. 8

Fig. 6. Oscilloscope traces of the transmitted signal received by a horn located behind the plasma layer: (a) incident power 0.25 MW; (b) power 1 MW. The time the plasma becomes transparent shifts to the right as the plasma density is increased.

Fig. 7. The change in the nature of the reflection from a dense plasma as the microwave power is increased: (a) incident wave field $E_0 = 80$ V/cm; (b) $E_0 = 800$ V/cm; (c) $E_0 = 1350$ V/cm. A signal of amplitude 5 divisions corresponds to $R = -1$; the detection is linear; $n_{max}/n_c = 4$.

Fig. 8. The low-frequency oscillations recorded by various instruments: (a) the microwave signal received by the waveguide (with and without plasma); (b) the ion current from a single Langmuir probe (with and without microwaves); (c) the ion saturation current at the multigrid probe; (d) diagnostic emission at a frequency (f_0 -900) MHz in the presence of high-power microwaves. The power of the incident wave is at least 100 kW.

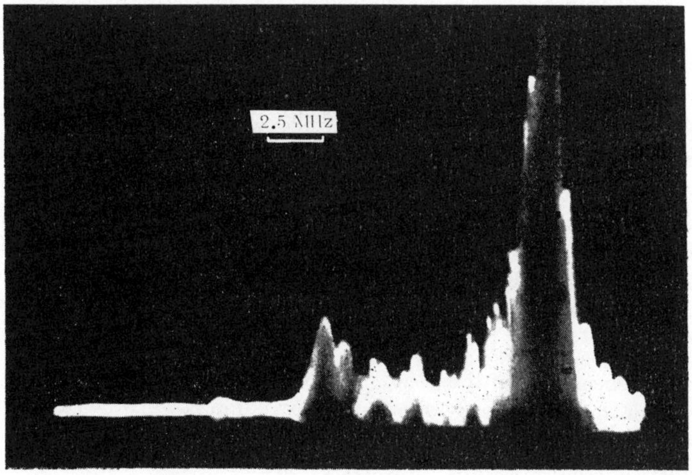

Fig. 9. The spectrum of the microwave radiation transmitted through a plasma layer. Incident power 1 MW; strobe time 2-4 μsec.

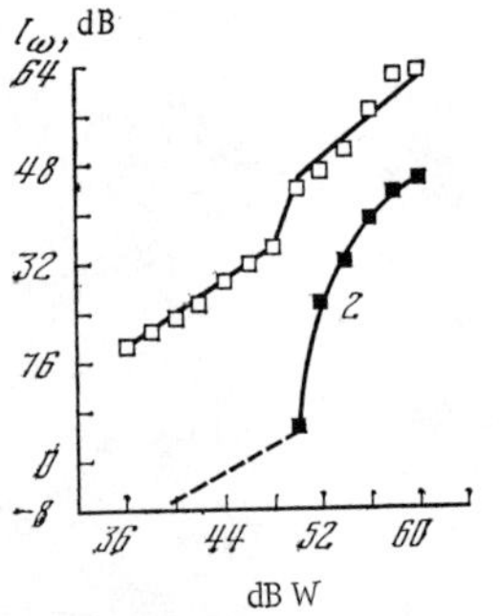

Fig. 10

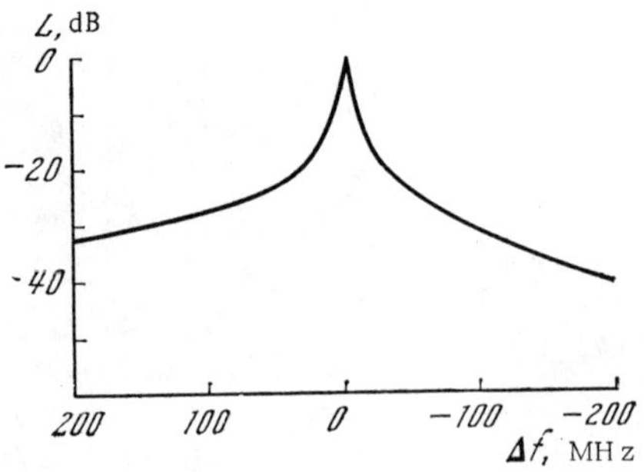

Fig. 11

Fig. 10. The dependence of the intensity of spectral components on the input power: (1) emission at frequency f_0; (2) emission at frequency ($f_0 - 10$) MHz. The dashed line corresponds to the level of reliable measurements.

Fig. 11. The attenuation in the supplementary filter as a function of the frequency.

energy and its dissipation) shows that the supplemental spectrum is caused by nonlinearity of the plasma, rather than by possible overloading of the crystal mixer at the input of the measurement circuit. The supplementary spectrum covers 10-12 MHz on the low side. No supplementary spectrum was observed on the high side to within the sensitivity of the measurement circuit (−30 dB). It should be noted that the amplitude scale of the spectral components in Fig. 9 is nonlinear. The measurements show that the maximum intensity of the displaced lines is 16 dB below f_0.

The dependence of the intensity of emission at the combination frequencies (Fig. 10) on the input power has a sharply nonlinear character and the intensity of the satellites reaches a level such that they can be detected with certainty only when $E_0/E_p = 0.2$. This value of the field corresponds to an input power level at which the layer becomes noticeably transparent within 2 μsec of the time the microwaves are turned on.

Analogous changes in the spectrum are recorded in the reflected signal; however, in this case the level of the satellites reaches −(10-12) dB of the fundamental. It is important to note that the threshold for detection of the satellites in this case is also the same as the threshold for transparency of the layer.

The spectra of the radiation which penetrated the plasma layer were also measured with the aid of a set of tunable measurement receivers (series P5) with a bandwidth of 4-5 MHz. In these measurements the question of removing the influence of the fundamental signal in detecting the satellites becomes important. Series P5 receivers have a built-in filter-preselector consisting of a resonator which is simultaneously tunable with the instrument. However, the decoupling provided by a single resonator is small (about 30 dB) and as the fundamental frequency is approached the decoupling decreases and may only be a few decibels at the edge of the passband. Thus, a VST-10P wavemeter resonator was inserted in the circuit to provide additional protection for the receiver from high-power radiation. The dependence of the additional decoupling on the frequency shift relative to the microwave oscillator frequency is plotted in Fig. 11.

The measurements made with the P5-5B detector (near f_0) confirm the spectral shape obtained with the single pulse analyzer. With time the combination radiation signals typically appear as several peaks which occupy the entire region from the time the microwave field is turned on until the plasma decays. Strictly speaking, these measurements do not allow us to follow the evolution of the supplementary spectrum. It can only be stated that the time for establishment of the spectrum does not exceed 1 μsec (it is affected by the temporal resolution and by the time for the pure spectrum of the oscillator to be established).

We now consider the dependence of the penetration of the signal through the layer on the incident power (see Fig. 10). An analysis of this dependence shows that for pump powers of 10-60 kW the power of the transmitted signal increases as the incident power to the power 1.3-1.5. When the incident power is increased from 60 to 100 kW the transmitted signal grows by an order of magnitude. As the incident power is increased further to 1 MW the transmitted signal increases as the square of the pump power. Therefore, over the entire range of incident powers the transmission is nonlinear and the nature of the nonlinearity changes with the appearance of "fast" transparency.

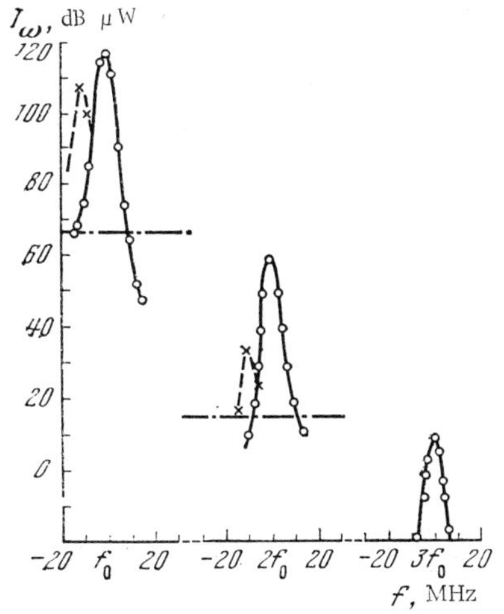

Fig. 12

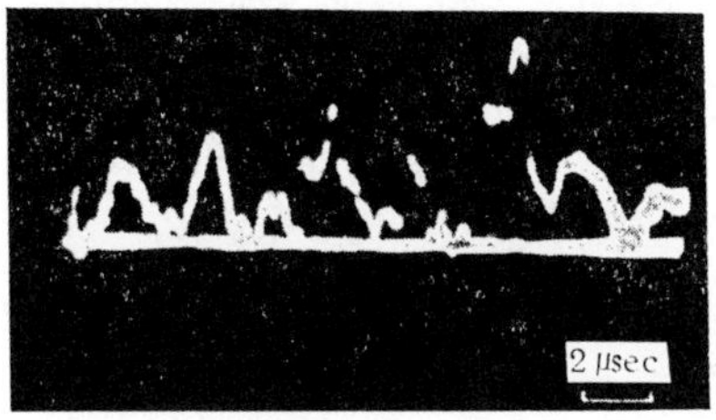

Fig. 13

Fig. 12. Spectra of the radiation entering the layer. The smooth curves correspond to the oscillator spectrum, and the dashed curves, to supplementary lines which appear when a plasma is present. Small changes in the intensity of the fundamental and its harmonics are not shown.

Fig. 13. Emission near the second harmonic of the oscillator. Power, 1 MW; shift from $2f_0$, 7 MHz.

7. Emission near Harmonics of the Pump Wave

Lengths of undersized waveguide were used as filters in the measurements of the upper harmonics of the oscillator emission and of the parts of the spectra near them. The attenuation at the fundamental frequency for such a waveguide is easily calculated and is given by [70]

$$L_f = 20 \log (e^{\beta z}),$$

where the damping constant $\beta = (2\pi/\lambda c)\sqrt{1-(\lambda_0/\lambda c)^2}$.

Lengths of 15 × 35 mm and 10 × 23 mm waveguide were employed. The attenuations were −300 dB for the second harmonic and −1000 dB for the third harmonic range.

These measurements confirmed that the level of harmonics in the oscillator with respect to the fundamental is −(60 ± 2) dB for the second harmonic and −(110 ± 2) dB for the third. The errors in these measurements are determined by the accuracy of measurements of the absolute value of the power using the receivers. Figure 12 shows the intensity of the spectral components of the emission as a function of the frequency in these bands. The horizontal dot-dashed lines denote the limits of reliable measurements. The widths of both the fundamental and harmonic lines do not characterize the emission spectrum but are determined by the bandpass of the detectors (4-5 MHz).

Repeated measurements in the presence of plasma show that the intensities of the second and third harmonics increase somewhat (by 3-5 dB) compared with the oscillator alone. This increase is recorded when there is an anomalous rise in the transparency of the dense plasma layer and can evidently be explained by focusing of the rays in the channel formed by the action of the pump wave.

Figure 12 also shows two values of the frequency at which peak intensities of the shifted (from the fundamental and harmonics of the oscillator) frequency are observed: $(f_0 - 10)$ MHz and $(2f_0 - 10)$ MHz. The intensity of the first line is 10 dB lower than f_0 and the second is 25 dB lower than the level of $2f_0$.

As an example we can follow the dependence of the line at $(2f_0 - 10)$ MHz on the power of the wave incident on the plasma. The line intensity rises with the power, reaches a maximum at a power of 600 kW, and then falls. No large changes in the frequency of the satellite with power were observed.

It is interesting to note that unlike the pulse of radiation at a frequency of $2f_0$, the shifted emission consists of isolated bursts (Fig. 13) similar to those observed at a frequency of $(f_0 - 10)$ MHz.

CHAPTER III

CURRENTS AND ENERGY SPECTRA OF FAST ELECTRONS PRODUCED IN THE PLASMA LAYER

Based on our observations of nonlinear transparency of the plasma layer and the detection of combination lines in the scattered spectrum, we may assume that a strongly turbulent state develops in the plasma layer such that the field strength in the electron plasma waves can reach values on the order of the characteristic plasma field E_p. We may thus expect substantial changes in the electron velocity distribution and, especially, intense heating of the electron component of the plasma together with the appearance of groups of fast electrons. It is clearly possible to estimate the intensity of dissipation of the electromagnetic waves from the enhancement in the energy content of the electrons and ions in the plasma. It is quite probable that the character of the wave processes which occur in the plasma is also reflected in the angular distribution of the charged particles, the time evolution of their distribution functions, and so on, as well as in their energy spectra. In addition, nonuniform heating of the electron gas and the appearance of a displacement wave [14, 15] or similar effects can lead to a redistribution of the plasma density, that is, to self-interaction of the incident wave. In this regard, it is of interest to study the time evolution of the electron currents and to study their spatial distribution, as well as to measure the increase in energy of the electron gas.

8. Features of Measurements in the Plasma Layer

The energy spectra and currents of the fast electrons were measured with the previously described multigrid probe located 18 cm from the aperture plane of the emitter.

Under these conditions the operation of the probe has a number of specific features. Because the plasma density is high it is necessary to apply high voltages to the decoupling grids. For a small probe this leads to electrical breakdown. Reducing the density of charged particles by means of input attenuating grids reduces the magnitudes of the detected currents and decreases the time resolution. Because of this condition, the energy of the electrons was analyzed before breakup of the plasma by varying the negative voltage on two grids which followed the input grid of the probe. Since the cell size in this grid is roughly equal to the Debye radius, the space in the plane of the grids could be regarded as equipotential. The probe collector and suppressor grid were at a high positive potential so that all the electron current which passed through the analyzer grids was collected.

In this operating regime, however, the probe cannot measure the angular distribution of the electron currents. This is because if the probe axis is parallel to the axis of the vacuum chamber, then the plasma ions do not enter the inlet aperture region of the probe since they move at almost right angles to the chamber axis from the plasma injectors. Thus, the ions do not compensate the space charge of the electrons and the magnitude of the electron current falls by more than an order of magnitude due to coulomb repulsion of the electrons. In order to eliminate errors in comparing currents in different directions, the experiment was set up so that ions did not enter the analyzer gap of the probe at all. This was possible because the plasma layer was created using only two injectors which aimed plasma in the x-direction while the probe could only detect particles moving in the y- and z-directions. Thus, the plasma had a directed velocity only in the direction perpendicular to the direction of the current that was being analyzed.

It should be noted that in experiments where the directed velocity of the ions does not allow them to enter the analyzer, there is a limit in the peak electron current due to space charge of the beam. In our case this limit can be described by Child's law for a plane gap (see, for example, [71]) which has the form (CGSE units)

$$i_{sat} = \frac{1}{9\pi}\left(\frac{2e}{m}\right)^{1/2}\frac{U_{sat}^{3/2}}{b^2},$$

where eU_{sat} is the energy of the electrons entering the gap and b is the length of the gap along their path. Applying voltages to the probe grids in this case would give the same results as in the previous case. The magnitude of the measured currents would reach 0.2-0.5 times the limiting values given by Child's law. Because of this, we can only derive very qualitative conclusions about the velocity distribution of the electrons from the dependence of the electron current on the retarding potential on the analyzer grids.

Another peculiarity of the probe measurements when the microwaves are interacting with the plasma is the appearance of a high positive plasma potential in the interaction region. Measurements of the floating potential of a single Langmuir probe with a load resistance on the order of the resistance of the plasma-probe

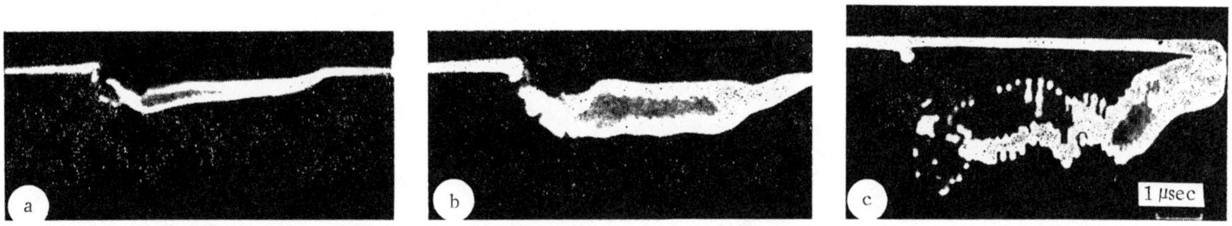

Fig. 14. Oscilloscope traces of the fast electron current taken with high time resolution: (a) incident wave field $E_0 = 1.0$ kV/cm; (b) 1.2 kV/cm; (c) 1.5 kV/cm.

interval showed that during the microwave pulse the probe potential rises by several tens of volts, reaching +(60-90) V with respect to the chamber walls. Clearly, neither fast electrons nor photoemission from the probe can be responsible for such high probe potentials since the energies of the bulk of the secondary and photoelectrons are at most a few tens of electron volts. In just the same way, fluxes of fast ions cannot raise the probe potential relative to the plasma potential because their current is compensated by the electron current. Thus, it can be stated that the positive potential of the plasma layer during the interaction with the microwaves rises relative to the vessel walls by at least the potential of the floating probe. This circumstance must be taken into account in evaluating the energy of the electrons detected by the multigrid probe since the potential of its case and first grid were equal to the potential of the vacuum chamber.

It is interesting to note that the floating potential of the probe is not uniformly distributed over the volume of the plasma layer. If the potential rises along the radius of the layer in the interaction region, then along the layer at medium powers (40-80 kW) there is a periodic variation in the potential with a characteristic length of 5-7 cm. An actual rise in the potential is detected only in the forward layers of the plasma and is damped toward the center of the plasma layer (z = 24 cm).

9. Fast Electron Currents, Time-Dependent Dynamics, and the Spatial Distribution

Fast electron currents were observed if the maximum electron density in the layer equaled or exceeded the critical density for the incident radiation. This phenomenon is characterized by a threshold power of the incident radiation of 100 kW which corresponds to a field strength in the incident wave at the probe location of 360 V/cm or to a ratio $E_0/E_p = 0.3$.

The probe collector was loaded with a resistance of 50 Ω, equal to the impedance of the cable connecting the collector to the amplifier input, in order to increase the time resolution. The upper frequency of the amplifier was 25 MHz. It was found that the electron current oscillates practically throughout the time it exists (Fig. 14). This time coincides with the time for the decay, caused by the high-power microwave field, of the plasma layer. The electron current appears in about 0.1 μsec which is equal to the risetime of the microwave pulse. This time is practically unchanged over the range of incident powers which corresponded to electric fields of from 360 to 1750 V/cm. It is interesting to note that the pattern observed in this case is different from that observed in [72] where the risetime of the probe current was several microseconds, comparable to the pairwise collision time. In our case we have risetimes for the electron currents that are two orders of magnitude shorter than the electron–ion collision time. It is also very interesting that the fall time of the current was not more than 0.1 μsec. (The fadeout time was measured in experiments where the microwave oscillator was turned off in the middle of the test interval. Then it was possible to obtain oscilloscope traces on which the electron current fell over a time less than the period of the oscillations in the electron current.) In this case our results are also different from those of [72, 73] where relaxation of the electron currents occurred over the pairwise collision time.

The dependence of the modulation depth (m%) of the electron current on the incident power was characterized by a sharp threshold, also equal to 100 kW, beyond which it remains at a level of about 40% (Fig. 15). At high powers the modulation depth has a slight tendency to decrease. An interesting feature of this process is the change in the frequency of the oscillations as the power level of the incident wave is changed. It follows from Fig. 14 that for a small power above threshold the frequency of the oscillations is much greater than 20 MHz and cannot even be measured with our apparatus (an S1-15/1 amplifier and S8-9A oscilloscope). As the excess above threshold is increased, the frequency of the oscillations decreases (Fig. 16) and varies roughly as $f_{osc} \sim \ln(E_{thr}^2/E_0^2)$. In our experiments frequencies from 22 MHz down to 4 MHz were recorded as the

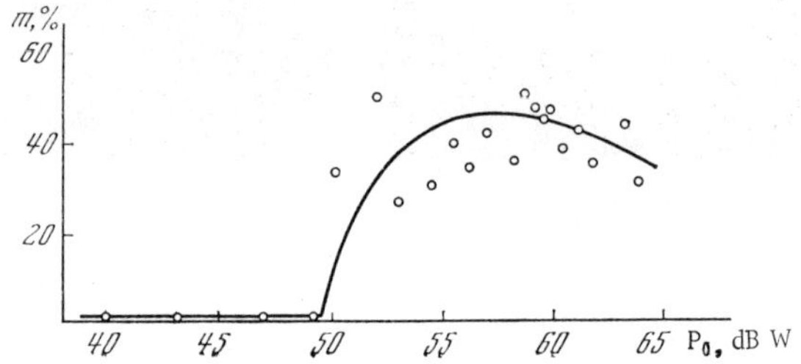

Fig. 15

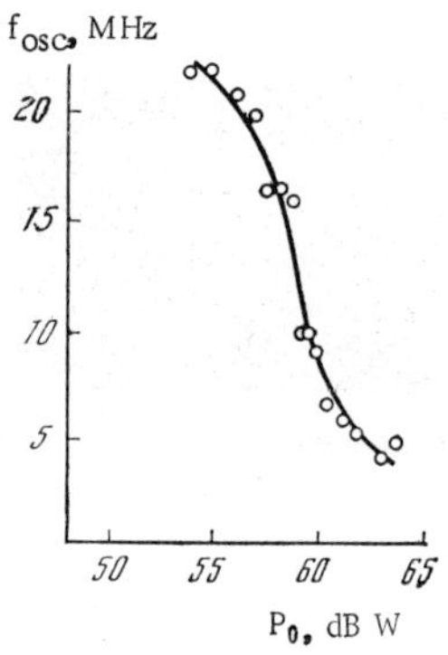

Fig. 16

Fig. 15. The modulation depth of the electron current as a function of the incident power. The scale on the abscissa is logarithmic. The threshold for appearance of oscillations is a field of 360 V/cm. The curve corresponds to times from 1.0-1.5 μsec after the incident wave has been turned on.

Fig. 16. The dependence of the frequency of the electron current oscillations on the incident wave power. The curve has been constructed for the time interval 1.0-1.5 μsec.

incident wave power was varied. It is interesting that the shifts in the rf spectrum recorded with the S4-14 spectrum analyzer fall in this interval.

In order to analyze our results it is also important to know the low-frequency spectra of these oscillations. The simplest information on the low-frequency spectrum can be obtained from an analysis of oscilloscope traces of the sort shown in Fig. 14. An analysis of the spectral content of these traces (Fig. 17) shows that frequencies higher than f_{Li} (approximately 70 MHz) do not appear in the electron signal. The peak in the spectral distribution occurs at 9 MHz.

The character of the oscillations does not change substantially if the retarding potential is raised all the way up to the maximum values at which fast electrons are still detected.

As can be seen from the traces in Fig. 14, in addition to these rapid oscillations, the electron current undergoes relatively slow changes in time. A clearer idea of these time variations may be obtained if the probe is located within the layers of dense plasma at the time the oscillator is turned on (Fig. 18). Since the collector was loaded with a resistance of 3 kΩ in this case, the rapid oscillations in the current were integrated and only the slow fluctuations could be detected (the time constant is 0.6 μsec). It is clear that, unlike in the previous case, where the electron density near the probe at the time the microwaves were turned on was close to critical, the current rise occurs over a period of several microseconds which evidently corresponds to a process of reducing the charged particle density near the probe. In addition, there are successive drops and rises in the current which can obviously be related to changes in the density of the plasma in the layer, in its potential relative to the probe, or in the conditions under which the electrons are heated. In every case the character of the oscillations in the electron current is identical to the fluctuations in the plasma density recorded by the microwave, Langmuir, and multigrid probes (Fig. 8). Therefore, this result indicates that the magnitude of the electron current is not uniform within the plasma layer but is probably a function of position. In this connection, it is important to know the relation of the size of the "hot" region to the size of the focal spot of the radiation.

This question was answered for electrons moving along the field by moving the multigrid probe in the x-direction. The size of the "hot" region (see Fig. 19) is equal to or even a bit smaller than the focal spot as determined by the half width of the field distribution.

10. Angular and Energy Distributions of the Fast Electrons

As noted above, in studying the energy spectra of the electrons it was appropriate to employ low time resolution. With a load resistance of $R_L = 3$ kΩ on the probe it was possible to detect electron currents with retarding potentials up to the order of 100 V and to detect current signals in various directions.

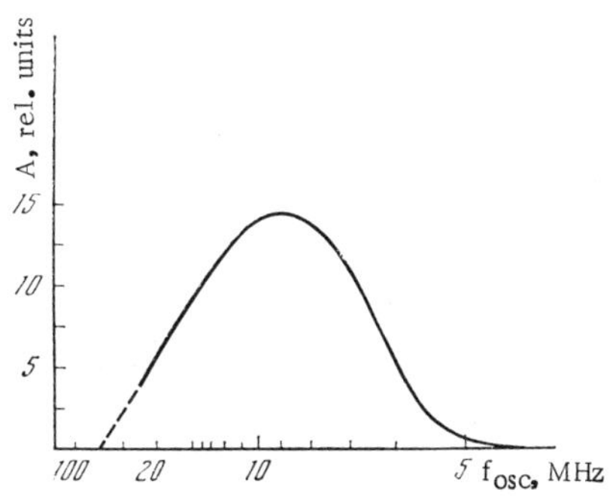

Fig. 17

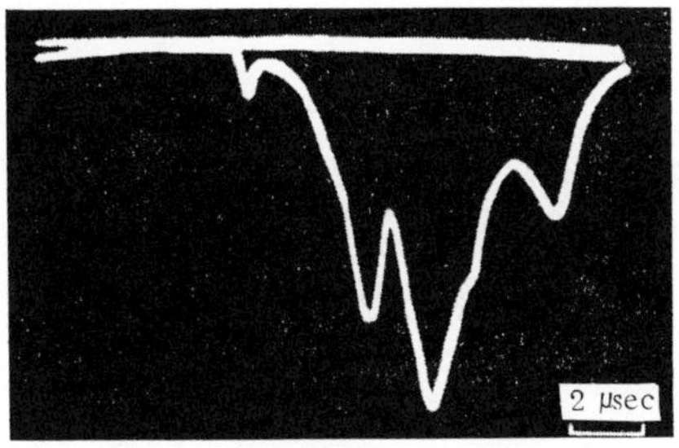

Fig. 18

Fig. 17. Analysis of the spectral content of an oscilloscope trace of the oscillations in the electron current (at the peak power of the incident wave).

Fig. 18. An oscilloscope trace of the multigrid probe current in a dense plasma with $n \simeq 2n_c$. The oscillator power is 200 kW; $U_{probe} = -10$ V.

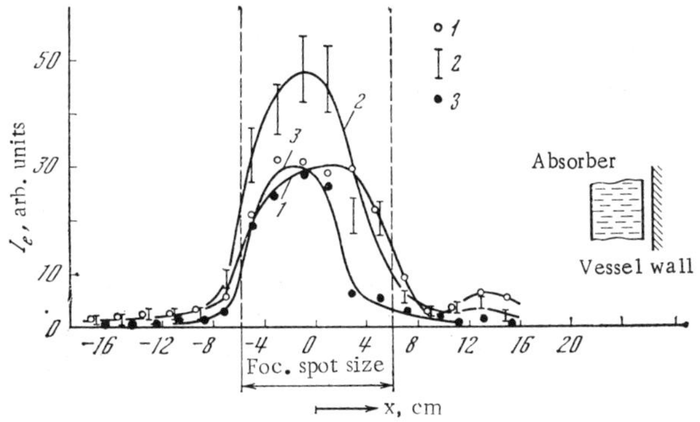

Fig. 19. The variation in the electron current picked up by the multigrid probe as it is moved transversely: (1), (2), and (3) correspond to times of 2, 4, and 6 μsec; power 500 kW; $n_{max}/n_c \simeq 3$.

Some previous experiments on this apparatus [11] showed that the energy of the bulk of the electrons increases by several times when strong microwaves act on the plasma. In addition, fast electrons, whose energies exceed the thermal energy by more than an order of magnitude, were detected. These results were obtained under conditions such that ions entered the analyzer gap of the probe (they were cut off only by the positive potential on the collector) and were compensated by the electron space charge. Here it was necessary to direct the probe toward one of the plasma injectors, and it was not possible to compare correctly the electron currents in different directions since when the probe was rotated the ions ceased to compensate the electron space charge and the electron current fell by roughly a factor of 10.

Later on, the fast electron component was the major topic of interest and it was studied under conditions such that the ions did not enter the probe at all. Figure 20 shows the correct ratio of the electron currents in different directions and the shape of their energy spectrum. The dependence of the electron current on the retarding potential for various probe orientations with an incident wave power of 500 kW is shown there. It should be noted that the energy of the electrons reaching the collector differs from the potential U plotted on the abscissa of Fig. 20 by an amount φ, the plasma potential. We note the slight divergence between the currents in the + and $-z$ directions (especially at the beginning of the curve). This divergence can be explained by the fact that when the probe is rotated by 180° its inlet aperture is shifted by roughly 3 cm in the coordinate system attached to the apparatus, that is, the probe lies in a slightly more rarefied plasma when it is measuring the electrons in the $-z$ direction. On the whole, we may conclude that the fast electron currents in various directions do not differ by more than 2-3 times.

These dependences of the currents on the retarding potential make it possible to qualitatively determine the changes in the electron distribution function by, for instance, graphical differentiation. The resulting distribution (Fig. 21) is the density of particles per unit velocity interval, that is, the velocity distribution. In

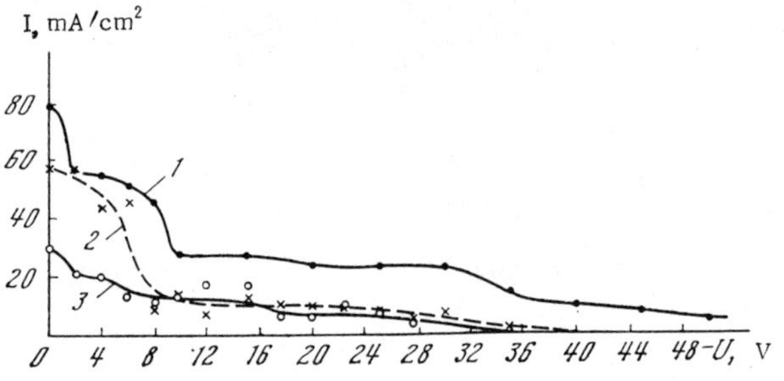

Fig. 20

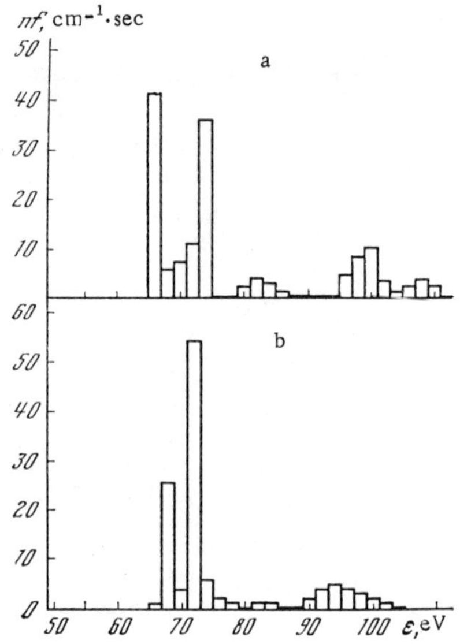

Fig. 21

Fig. 20. The dependence of the electron current on the retarding potential for various probe orientations: in the y-direction (1); +z-direction (2); −z-direction (3); $n_{max}/n_c \simeq 3$.

Fig. 21. Histograms of the electron distribution as a function of energy: (a) y-direction; (b) z-direction.

constructing Fig. 21 we have taken the floating potential to be 50 eV, i.e., the histograms are shifted toward higher electron energies. Thus, they reflect only the fast electron "tail." From the figure we may conclude that there are groups of electrons with energies of about 60 and 85 eV when the maximum energy of the oscillations $e^2E_0^2/2m\omega_0^2 = 0.8$ eV.

11. Evaluation of the Level of Langmuir Noise in the Plasma Layer

Measurements of the currents and energies of the electrons with a multigrid probe make it possible to evaluate the absorption coefficient for the pump wave energy and the intensity of Langmuir noise in the plasma. By computing and summing the fluxes of electron energy in different directions at a point it is possible to determine the total energy flux in the plasma through unit area. By relating this quantity to the energy flux of the incident radiation, we obtain the absorption coefficient. If we assume that the absorbed microwave energy is first transformed into Langmuir waves and only later into the energy of fast electrons, then, we may obviously assume that the energy flux in Langmuir waves is at least as large as the energy flux in the fast electrons. Equating these two quantities, we easily obtain an estimate of the energy density of the Langmuir noise:

$$kr_{De}\,v_{Te}\,E_l^2/8\pi \simeq n_f\bar{v}_f\,\bar{\varepsilon}_f.$$

Here n_f, $\bar{v}_f$, and $\bar{\varepsilon}_f$ denote the density, mean velocity, and mean energy in the fast electron flux.

An estimate of the energy flux carried by the fast plasma electrons alone may be obtained by integrating the multigrid probe current as a function of the retarding potential.

In doing this it is necessary to include the additional energy lost by the electrons as they go to the walls from the positively charged plasma column.

An analysis of the volt–ampere characteristics obtained at different probe positions shows that for an oscillator power of 1.25 MW roughly 3% of the energy flux of the incident wave goes into heating of electrons parallel and antiparallel to the **E** vector and roughly 1% into heating of electrons parallel and antiparallel to the wave vector of the pump wave. If we assume that similar energy fluxes occur along a third axis perpendicular to the field and wave vectors, then the absorption coefficient reaches 6%. Thus the average energy density of the Langmuir noise is $E_l^2/8\pi nT_0 \simeq 10$. This estimate shows that we are dealing with a highly turbulent layer of plasma in which the electron density must evidently be close to zero at the places where the

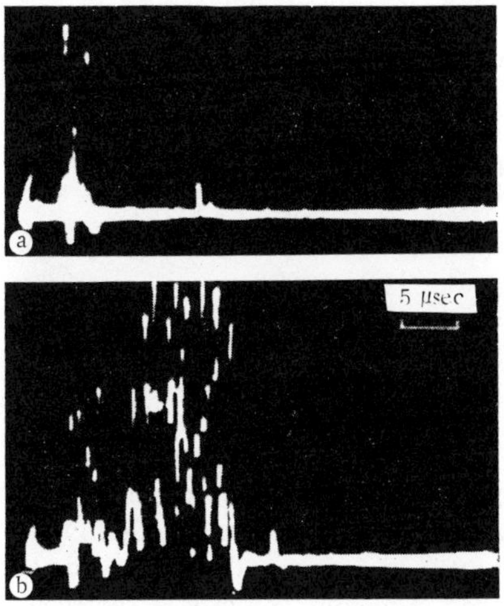

Fig. 22. Plasma emission signals picked up by the measurement receiver (extra filtering has been used to protect it from the high-power microwaves): (a) intrinsic plasma emission without the high-power oscillator; (b) plasma emission in the presence of microwaves at a power of 100 kW. Shift from the fundamental is 900 MHz.

Langmuir waves are localized. A plasma layer with such modulations in the electron density may appear transparent to the pump wave.

12. A Study of Electron "Heating" in Terms of the Microwave Emission from the Plasma

In a collisionless plasma, when the mean free path of the electrons exceeds the size of the plasma layer, the intensity of the thermal radiation at frequencies above the plasma frequency is extremely low because of the small absorption coefficient. However, if the frequency lies below the plasma frequency corresponding to the maximum electron density in the layer, then because of linear transformation the emission is close to that from an absolute blackbody. It has been shown [74] that the intensity of the emission in this case is given by $I = 0.5AI_0$. Here A is the transformation coefficient, equal to 0.4 for the optimum angle [20], and I_0 is the intensity of the emission from an absolute blackbody. If the radiation is detected with the aid of a single mode waveguide, the received power is equal to

$$P \simeq {}^1/_2 A T_r \Delta f,$$

where T_r is the radiation temperature of the plasma and Δf is the receiver bandwidth. It is interesting to note that if linear transformation is neglected, then the range of frequencies in a collisionless plasma where its absorption coefficient is close to that of an absolute blackbody may be considerably narrower than the receiver bandwidth [75]. Linear transformation causes the absorption coefficient to be almost constant over a wide band and the width of the absorption region is greater than the receiver bandwidth.

The thermal emission from the plasma was picked up by a magnetic loop mounted in a pipe and oriented at roughly 45° to the surface of the plasma layer. The loop was coupled to a resonator by a length of coaxial cable and the signal from the resonator entered a P5-4 measurement receiver with a bandwidth of 4.5 MHz and a sensitivity of about 10^{-11} W.

A shift of 900 MHz from the oscillator frequency was used to suppress signals from the pump wave generator. Consequently, emission from the region $n \simeq 0.5n_c$ was recorded. Oscilloscope traces of the signal are shown in Fig. 22. When the pump wave is not turned on the signal is well above the noise level of the receiver at the moment the plasma guns are turned on. In addition there are bursts of radiation which exceed the noise by 2-4 times in the interval between 5 and 10 μsec, that is, at the time the plasma flows from the various injectors interact. Evidently at these times the electron component of the plasma is being heated by slowing down of the plasma flows through collective interactions. An estimate of the electron temperature at this time yields several tens of electron volts (the radiation power is 10^{-11} W), and at the time of the discharge in the injectors, several hundred electron volts. Turning on the microwaves leads to a rise in the intensity of the emission by more than an order of magnitude ($\sim 10^{-10}$ W) which can be attributed to a rise in the average energy of the electrons to several hundred electron volts. It is interesting to note that the time variation of the radiant intensity in all the cases that were examined corresponds to short isolated bursts lasting a few tenths of a microsecond each.

A comparison shows that these results differ in order of magnitude from the results of the probe measurements. Evidently the microwave emission from the plasma is not thermal but is caused by beams of electrons propagating into a region of rarefied plasma ($n \simeq 0.5n_c$) from denser layers.

We therefore have yet another indication of the non-Maxwellian character of the electron velocity distribution.

CHAPTER IV

DISCUSSION OF THE EXPERIMENTAL RESULTS

Several important experimental facts must be noted before proceeding to a discussion of the results of these measurements.

If we examine the dependence of the intensity of the transmitted wave on the intensity of the incident wave, we find that at around 100 kW there is a sharp enhancement in wave penetration through the layer. This effect is seen even in the front of the pulse; that is, it develops over a period of less than 0.1-0.2 μsec. The level of the signal that penetrates the plasma layer is 0.1-0.2 times the incident signal level. This value varies little for 2-3 μsec after the process begins. Over this same period strong oscillations in the plasma density are observed by means of the microwave, Langmuir, and multigrid probes. The further development of the process involves increased levels of the transmitted signal and is accompanied by a reduction in the charged particle density in the plasma layer.

Besides increased transparency of the plasma layer as the power is increased to 100 kW, three other processes are observed: generation of red satellites in the spectra of the radiation scattered and transmitted through the layer, excitation of radiation near the second harmonic of the pump wave, and production of fluxes of fast electrons. All three processes occur for several microseconds before the charged particle density in the plasma layer has fallen. We note that the fast electron currents and the intensity of the emission near the second harmonic oscillate in time. In addition, as we have seen, these phenomena are accompanied by intense transfer of energy from the incident wave to the electron component of the plasma.

We shall now attempt to compare these facts with various nonlinear processes. We first consider the phenomenon of "fast" transparency. It is easy to see that the breakup of the layer into bunches of plasma of size on the order of the wavelength along the propagation direction of the wave or the formation of a channel in the depth of the layer can occur over times of order $\tau = \lambda_0/v$, where v is either the sound speed or the speed of the plasma in the microwave quasipotential $\Phi = e^2E_0^2/4m\omega_0^2$ ($v \sim \sqrt{e\Phi/M}$). Since $E_0 > E_p$, an estimate based on the quasipotential yields a lower value of the speed. Thus, it appears that $\tau = 3$-5 μsec, which is an order of magnitude greater than the experimentally recorded time for "fast" transparency to occur.

We are therefore apparently dealing with different mechanisms for the transparency than those discussed in [5-8, 60-67].

Rf pressure forces in the region where the field is being amplified must lead to more rapid redistribution of the plasma density. When $E_0/E_p \simeq 0.3$ (power 100 kW) field amplification at the singular point is already limited by nonlinearity in the electron motion [24]. In this case for the optimum angle of incidence we have $E \simeq 25E_p$ and the transparency time for the entire thickness of the plasma layer ($\sim 3\lambda_0$) is about 0.5 μsec or on the order of magnitude of the experimental value. However, it should be noted that the angle of incidence of the waves is far from optimal and the experimental geometry is such that on the axis of the layer the axial component of the field is attenuated by defocusing of the wave. It should also be added that as the ions are displaced from the singular point ($n \simeq n_c$) there is a substantial reduction in the field strength [24].

The assumption that "fast" transparency occurs due to striction forces in the field amplification region contradicts the threshold character of the transparency when $E_0/E_p \simeq 0.3$ since field amplification occurs at arbitrary pump wave intensities and values of $E/E_p = 25$ are reached at powers of 20-30 kW when the spatial dispersion is limited.

We have suggested previously [10] that as a result of a parametric instability, a modulation is excited in the plasma density along the electric field of the wave when there is an instability in an intense acoustic wave. The modulation of the plasma density along the field of a pump wave because of rf pressure forces in a standing Langmuir wave was examined in [68]. Estimates of the transparency time given there yield a value of 5×10^{-8} sec, which agrees in order of magnitude with the experimental data. We thus arrive at a need to include the idea of parametric excitation of waves in order to explain the "fast" transparency effect.

Besides "fast" transparency, at fields $E_0/E_p \geq 0.3$ fast electrons are produced. This effect can apparently also be explained by parametric excitation of Langmuir plasma turbulence. It would seem that "heating" of the electron component can also be a consequence of the linear transformation of waves [18]. However, in this case there are so threshold fields and the electrons must gain energy along the density gradient if the Langmuir waves are damped by Landau damping rather than by collisions. The pattern we have observed does not fit into this framework. The same threshold intensity of the pump wave occurs as for the "fast" transparency while the fast electrons move both along the density gradient and perpendicular to it along the electric field of the microwaves.

If we now estimate the threshold fields for parametric excitation of oscillations in an inhomogeneous plasma (see Table 1), it is easy to verify that the threshold fields for "fast" transparency, combination frequencies in the emission spectrum,* and generation of electrons are at least an order of magnitude greater than the threshold fields for the four nonlinear processes.

According to [52] for the modulational instability we have

$$E_0/E_p > 2r_{De}/L = 10^{-3}.$$

Such a low value of the threshold field has apparently been confirmed experimentally in [72, 76, 77]. For decay of the pump wave into an ion acoustic wave and a Langmuir wave we have [38]

$$\frac{E_0}{E_p} > \left(8\pi\frac{m}{M}\right)^{1/4}\left(\frac{1}{k_{\parallel}L}\right)^{1/2} \simeq \left(8\pi\frac{m}{M}\right)^{1/4}\left(\frac{3r_{De}}{L}\right)^{1/2} = 10^{-2}.$$

This formula was verified experimentally by Eubank [78].

In [38] the following expression was obtained for modified decay:

$$\frac{E_0}{E_p} > \left(\frac{2}{k_{\parallel}L}\right)^{1/2} \simeq \left(\frac{6r_{De}}{L}\right)^{1/2} = 5\cdot 10^{-2}.$$

A value close to this is obtained for stimulated Mandelshtam–Brillouin scattering in inhomogeneous plasmas [39].

Estimates of the threshold fields for other nonlinear processes (see Table 1) yield values $E_0/E_p > 1$. Therefore, under our conditions even before "fast" transparency and combination scattering appear it is possible for at least these four nonlinear processes to occur. Apparently this means that one process can affect another. In this regard it is interesting to compare the maximum growth rates for various instabilities. Thus, for the modulational [50] and aperiodic [34] instabilities we have the same growth rate:

$$\frac{\gamma_{max}}{\omega_{Le}} \simeq \frac{1}{\sqrt{3}}\left(\frac{m}{M}\frac{E_0}{E_p}\right)^{1/2}.$$

For decay processes (periodic instability) we have [34]

$$\frac{\gamma_{max}}{\omega_{Le}} \simeq 0.34\left(k_{max}r_{De}\frac{m}{M}\frac{E_0}{E_p}\right)^{2/3}.$$

And, finally, for stimulated Mandelshtam–Brillouin (Brillouin) scattering we have [39]

$$\frac{\gamma}{\omega_{Le}} \simeq \frac{E_0}{E_p}\left(\frac{v_{Te}}{c}\right)^{1/2}.$$

A comparison of these expressions shows that the modulational and aperiodic instabilities have the fastest growth rates. It may be assumed that this fact is decisive in the competition among these processes. Perturbations along the density gradient must develop preferentially because of the field amplification effect near $n \simeq n_c$. The development of these perturbations leads, as shown experimentally in [72], to strong density perturbations in the same direction. The thresholds for parametric excitation of decay or quasidecay processes may then increase rapidly since the characteristic size of the inhomogeneity is evidently on the order of the characteristic size of the field amplification region, $(r_{De}^2 L)^{1/3}$ or $(r_E L)^{1/2}$ [24]. This may occur if the rate of change in the plasma density caused by rf pressure forces is comparable to the rate of development of the instability. The time for density perturbations to develop is roughly $(\omega_{Le}\tau)^{-1} \simeq \sqrt{\frac{m}{M}\frac{E_0}{E_p}}\left(\frac{r_{De}}{L}\right)^{-1/3}$ if $(r_{De}/L)^{2/3} > (r_E/L)^{1/2}$, and $(\omega_{Le}\tau)^{-1} \simeq (m/M)^{1/2}$ if $(r_{De}/L)^{2/3} < (r_E/L)^{1/2}$. It is easy to see that these times are, in fact, on the order of the reciprocal of the growth rate of the aperiodic instability. Evidently the plasma density oscillations observed at early times after the pump wave is turned on are caused by this effect.

---

*It is interesting to note that the threshold field we have measured for the excitation of the red satellite is more than an order of magnitude greater than the pump field used in Stern and Tzoar's experiment [76].

It is interesting to note that as the field is increased, when the nonlinearity begins to predominate in limiting the field at the singular point, the size of the amplification region increases and, therefore, the threshold for parametric excitation of waves perpendicular to the density gradient is reduced.

If we now assume that the axial inhomogeneity in the plasma layer is of a size on the order of the field amplification region when it is limited by the nonlinearity, then we have the following threshold field for decay into an ion-acoustic and Langmuir wave:

$$\frac{E_0}{E_p} \geqslant \left(72 \frac{m}{M} \frac{r_{De}}{L}\right)^{1/5} \simeq 0.15.$$

It is thus apparently possible to suppose that only at such high pump wave intensities do Langmuir oscillations begin to develop along the field vector of the incident wave and, besides the oscillations whose frequency is essentially equal to that of the pump wave, frequencies are excited which are shifted relative to the pump by the acoustic frequency.*

The angular distribution of the fast electron currents can evidently serve as a confirmation of this hypothesis. In fact, the currents along the external electric field are even slightly larger than those along the density gradient of the plasma. It is also important that the threshold fields for excitation of the fast electron currents are the same as those for occurrence of "fast" transparency and combination frequencies. The very fact that fast electrons appear is also in accordance with the assumption that the periodic instability is excited. In this case short wavelength Langmuir oscillations develop with kr_{De} = 0.2-0.3 and whose Landau damping decrement is

$$\frac{\gamma_L}{\omega} = \sqrt{\frac{\pi}{8}} \frac{1}{(kr_{De})^3} \exp\left(-\frac{1}{2k^2 r_{De}^2}\right) = 3 \cdot 10^{-4} \text{ to } 10^{-1},$$

so that energy is rapidly transferred from the waves to the electrons. That this is so is indicated by the rapid fall in the electron current after the pump wave is turned off. As we have already noted, this result is qualitatively different from the results of [72, 73], where the damping time for the waves and for the perturbations in the electron current are of the same order of magnitude as the pairwise collision time.

We now consider the transient nature of the processes taking place in the plasma layer. The fact that they are nonstationary shows up most clearly in the oscillations in the fast electron current. But we have the same sort of oscillations in the nonthermal plasma emission and in the plasma radiation at the second harmonic of the pump wave. If the fluctuations in the nonthermal emission indicate that the electron currents are nonstationary, then those in the second harmonic emission are evidence of fluctuations in the Langmuir plasma noise. It may be assumed that the oscillations in the electron currents are also caused by fluctuations in the plasma noise level as has been proposed for the intensity of scattered pump light or x-ray bremsstrahlung in laser plasmas [3, 79]. However, unlike the work on laser plasmas, where the main wave excitation process is decay of a transverse wave into two plasmons, we are dealing with decay into an ion-acoustic and Langmuir wave. Because of this fact, the mechanism for the nonstationarity may be stimulated scattering of ion-acoustic waves on ions [48]. An estimate of the period of the oscillations according to Eq. (15) of [48] yields

$$T \sim 10^2 \frac{M}{m} \sqrt{\frac{T_i}{T_e}} \left(\frac{E_0^2}{E_{0t}^2} - 1\right)^{-1} \simeq 10^{-6} \text{ sec.}$$

Thus, the period of the oscillations is an order of magnitude less than that measured experimentally. Furthermore, there is a qualitative difference in the dependence of the frequency of the oscillations on the external field strength. Experimentally we have a reduction in the frequency as the field is raised, while the calculation shows that the frequency increases. However, we must remember that the calculation is valid near the threshold fields, while the measurements are made well above threshold and at such a high noise intensity that the expressions used in [48] are hardly applicable.

It is known that the nonstationarity of Langmuir turbulence is caused by induced scattering of Langmuir waves on ions in a number of cases [41-45]. Calculations of the period of the fluctuations yield a value on the order of 10 times the reciprocal of the maximum growth rate of the instability (see, for example, [43]). Evidently, the growth rate of the instability does not exceed the period of the ion Langmuir (plasma) oscillations. Thus, the order of magnitude of the period of the oscillations is $T < 10/\omega_{Li} \simeq (2\text{-}5) \cdot 10^{-8}$ sec, in agreement with the experimental data.

*The magnitude of the frequency shift makes it possible to estimate the magnitude of the wave vectors of the excited oscillations. In fact, $\Delta\omega = \omega_s \simeq kr_{De}(m/M)^{1/2}\omega_0$. Thus, $kr_{De} \simeq 0.15$.

CONCLUSIONS

As we have seen, the action of an intense electromagnetic wave on a layer of dense collisionless plasma leads to "fast" transparency of the plasma layer, intense fluctuations in the density, the appearance of combination frequencies in the spectrum of the scattered radiation, and the production of fast electron currents. These effects are characterized by relatively high threshold fields $E_0/E_p \simeq 0.3$ and, in a number of cases, by transient time behavior (oscillations in the intensity of the fast electrons, in the nonthermal emission from the plasma, and in the emission near the second harmonic). An analysis of the experimental data allows us to conclude that intense Langmuir turbulence exists in the plasma layer and that various nonlinear processes for parametric excitation of waves interact.

LITERATURE CITED

1. R. Z. Sagdeev, Usp. Fiz. Nauk, 110:437 (1973).
2. A. V. Gurevich, Usp. Fiz. Nauk, 117:184 (1975).
3. N. G. Basov, O. N. Krokhin, V. V. Pustovalov, et al., Zh. Eksp. Teor. Fiz., 67:118 (1974).
4. G. M. Batanov and M. S. Rabinovich, V International Conf. on Plasma Physics and Controlled Nuclear Fusion, Tokyo (1974), Vol. 2, p. 625.
5. G. A. Askar'yan, Zh. Éksp. Teor. Fiz., 42:1567 (1962).
6. V. I. Talanov, Izv. Vyssh. Uchebn. Zaved., Radiofiz., 7:564 (1964).
7. A. V. Gurevich, Zh. Éksp. Teor. Fiz., 48:701 (1965).
8. V. P. Silin, Zh. Éksp. Teor. Fiz., 53:1662 (1967).
9. Yu. Ya. Brodskii, B. G. Eremin, A. G. Litvak, and Yu. A. Sakhonchik, Pis'ma Zh. Éksp. Teor. Fiz., 13:163 (1971).
10. G. M. Batanov and V. A. Silin, Pis'ma Zh. Éksp. Teor. Fiz., 14:445 (1971).
11. G. M. Batanov and V. A. Silin, Tr. Fiz. Inst. Akad. Nauk SSSR, 73:87 (1974).
12. Yu. Ja. Brodskii, S. V. Golubev, V. L. Gol'tsman, and A. G. Litvak, VI European Conf. on Controlled Fusion and Plasma Physics, Moscow (1973), Vol. 1, p. 549.
13. G. M. Batanov and V. A. Silin, Pis'ma Zh. Éksp. Teor. Fiz., 19:621 (1974).
14. A. A. Ivanov, L. P. Kozorovitski, and V. D. Rusanov, Dokl. Akad. Nauk SSSR, 184:811 (1969).
15. A. A. Ivanov, V. D. Rusanov, and R. Z. Sagdeev, Pis'ma Zh. Éksp. Teor. Fiz., 12:29 (1970).
16. N. G. Denisov, Zh. Éksp. Teor. Fiz., 31:609 (1956); 34:528 (1958): Radiotekh. Élektron., 1:732 (1956).
17. V. V. Zheleznyakov, Radiotekh. Élektron., 1:840 (1956).
18. V. E. Golant and A. D. Piliya, Usp. Fiz. Nauk, 104:413 (1971).
19. N. S. Erokhin and S. S. Moiseev, Usp. Fiz. Nauk, 109:225 (1973).
20. A. D. Piliya, Zh. Tekh. Fiz., 36:818 (1966).
21. A. A. Vedenov, Reviews of Plasma Physics, Vol. 3, Consultants Bureau, New York (1967).
22. V. B. Gil'denburg, Zh. Éksp. Teor. Fiz., 46:2156 (1964).
23. V. B. Gil'denburg and G. M. Fraiman, Zh. Éksp. Teor. Fiz., 69:1600 (1975).
24. S. V. Bulanov and L. M. Kovrizhnykh, Fiz. Plazmy, 2:105 (1976).
25. A. S. Kompaneets, Zh. Éksp. Teor. Fiz., 14:171 (1944).
26. V. L. Ginzburg and V. V. Zheleznyakov, Astron. Zh., 35:694 (1958).
27. A. I. Akiezer, I. G. Prokhoda, and A. G. Sitenko, Zh. Éksp. Teor. Fiz., 33:750 (1957).
28. T. F. Volkov, Plasma Physics and the Problem of Controlled Thermonuclear Reactions [in Russian], Vol. 4, Izd. AN SSSR, Moscow (1958), p. 98.
29. R. A. Sturrock, Phys. Rev., 112:1488 (1959).
30. V. N. Oraevskii and R. Z. Sagdeev, Zh. Tekh. Fiz., 32:1291 (1962).
31. V. P. Silin, Zh. Éksp. Teor. Fiz., 48:1679 (1965).
32. V. N. Oraevskii, Nuclear Fusion, 4:263 (1964).
33. L. M. Gorbunov, Usp. Fiz. Nauk, 109:631 (1973).
34. V. P. Silin, The Parametric Interaction of High-Power Radiation with Plasmas [in Russian], Nauka, Moscow (1973).
35. A. A. Galeev and R. Z. Sagdeev, Reviews of Plasma Physics, Vol. 7, Consultants Bureau, New York (1978); Lectures on Nonlinear Plasma Theory, ICTP, Trieste (1966).
36. K. J. Nishikawa, J. Phys. Soc. Jpn., 24:916; 1152 (1968).
37. A. D. Pilija, X Internat. Conf. on Phenomena in Ionized Gases, Oxford (1971), p. 320.
38. F. W. Perkins and J. Flick, Phys. Fluids, 14:2012 (1971).
39. C. S. Liu, M. N. Rosenbluth, and R. B. White, V Internat. Conf. on Plasma Phys. and Controlled Nuclear Fusion Research, Tokyo (1974), IAEA-CN-33/F5-2.

40. W. L. Kruer, P. R. Kaw, J. M. Dawson, and C. Oberman, Phys. Rev. Lett., 24:987 (1970).
41. E. J. Valeo, C. Oberman, and F. W. Perkins, Phys. Rev. Lett., 28:340 (1972).
42. V. E. Zakharov, S. L. Musher, and A. M. Rubenchik, Pis'ma Zh. Éksp. Teor. Fiz., 19:249 (1974); Zh. Éksp. Teor. Fiz., 69:155 (1975).
43. S. L. Musher and A. M. Rubenchik, Fiz. Plazmy, 1:982 (1975).
44. W. L. Kruer and E. J. Valeo, Phys. Fluids, 16:675 (1973).
45. N. E. Andreev, V. V. Pustovalov, V. P. Silin, and V. T. Tikhonchuk, Pis'ma Zh. Éksp. Teor. Fiz., 18:624 (1973); Kvantovaya Élektron., No. 5, 1099 (1974).
46. A. G. Litvak, V. Yu. Trakhtengertz, T. N. Fedoseeva, and G. M. Fraiman, Pis'ma Zh. Éksp. Teor. Fiz., 20:544 (1974).
47. I. A. Kol'chugina, A. G. Litvak, and I. V. Khazanov, NIRFI Preprint No. 56 (1974).
48. V. V. Pustovalov and V. P. Silin, Zh. Éksp. Teor. Fiz., 45:2472 (1975).
49. S. V. Kiyashko, V. V. Papko, and M. I. Rabinovich, Fiz. Plazmy, 1:1013 (1975).
50. V. E. Zakharov, Zh. Éksp. Teor. Fiz., 62:1745 (1972).
51. A. A. Vedenov and L. I. Rudakov, Dokl. Akad. Nauk SSSR, 159:767 (1964).
52. A. A. Galeev, R. Z. Sagdeev, V. D. Shapiro, and V. I. Shevchenko, Pis'ma Zh. Éksp. Teor. Fiz., 21:539 (1975).
53. V. B. Gil'denburg, Zh. Éksp. Teor. Fiz., 45:1978 (1963); Zh. Tekh. Fiz., 34:372 (1964).
54. H. H. Klein and W. M. Manheimer, Phys. Rev. Lett., 33:953 (1974).
55. Ya. N. Istomin, V. I. Karpman, and D. R. Shklyar, Fiz. Plazmy, 2:121 (1976).
56. L. M. Kovrizhnykh and A. S. Sakharov, Fiz. Plazmy, 2:97 (1976).
57. R. Y. Chiao, E. Garmire, and C. H. Townes, Phys. Rev. Lett., 13:479 (1964).
58. V. L. Ginzburg and A. V. Gurevich, Usp. Fiz. Nauk, 70:201; 393 (1960).
59. A. G. Litvak, Izv. Vyssh. Uchebn. Zaved., Radiofiz., 9:629; 675 (1966).
60. A. G. Litvak and G. M. Fraiman, Zh. Éksp. Teor. Fiz., 68:1288 (1975).
61. T. F. Volkov, Plasma Physics and the Problem of Controlled Thermonuclear Reactions [in Russian], Vol. 3, Izd. AN SSSR, Moscow (1958), p. 336.
62. F. G. Bass and Yu. G. Gurevich, Usp. Fiz. Nauk, 103:447 (1971).
63. R. Z. Sagdeev and V. D. Shapiro, Zh. Éksp. Teor. Fiz., 66:1651 (1974).
64. V. A. Mironov, Izv. Vyssh. Uchebn. Zaved., Radiofiz., 14:1450 (1971).
65. V. B. Gil'denburg, NIRFI Preprint No. 57 (1974).
66. K. Baumgärtel and K. Sauer, ZIE-Bericht 75-3, Akad. Wiss. DDR, Zentralinst. Elektronenphysik.
67. V. I. Karpman, Nonlinear Waves in Dispersive Media [in Russian], Nauka, Moscow (1973).
68. A. G. Litvak, V. A. Mironov, and G. M. Fraiman, Pis'ma Zh. Éksp. Teor. Fiz., 22:368 (1975).
69. W. H. Bostick, Phys. Rev., 106:404 (1957).
70. I. V. Lebedev, Microwave Technology and Devices [in Russian], Vol. 1, Gosénergoizdat, Moscow–Leningrad (1961).
71. W. Smythe, Static and Dynamic Electricity, McGraw-Hill, New York (1950).
72. A. Y. Wong and R. L. Stenzel, Phys. Rev. Lett., 34:727 (1975).
73. H. C. Kim, R. L. Stenzel, and A. Y. Wong, Phys. Rev. Lett., 33:886 (1974).
74. A. D. Piliya, Zh. Tekh. Fiz., 36:2195 (1966).
75. V. E. Colant, Microwave Techniques for Plasma Studies [in Russian], Nauka, Moscow (1962), p. 251.
76. R. A. Stern and N. Tzoar, Phys. Rev. Lett., 17:903 (1966).
77. V. A. Silin, Tr. Fiz. Inst. Akad. Nauk SSSR, 92:70 (1977), p. 70.
78. H. P. Eubank, Phys. Fluids, 14:2551 (1971).
79. Yu. S. Kas'yanov, V. V. Korobkin, P. P. Pashinin, A. M. Prokhorov, V. K. Chevokin, and M. Ya. Shchelev, Pis'ma Zh. Éksp. Teor. Fiz., 20:719 (1974).

COLLISIONLESS ABSORPTION OF ELECTROMAGNETIC WAVES IN PLASMAS AND "SLOW" NONLINEAR PHENOMENA

V. I. Barinov, I. R. Gekker,
V. A. Ivanov, and D. M. Karfidov

An experimental study is made of the interaction of pulsed microwaves in the 10 cm range with collisionless plasma flows ($\omega \gg \nu$) in a waveguide under both favorable ($\nabla n \parallel \mathbf{E}$, $\omega \approx \omega_{Le}$) and unfavorable initial conditions for plasma resonance of the waves. It is shown that in the first case over the interval $v_E/v_{Te} = 10^{-5}$–5 the absorption coefficient D^2 changes little while absorption sets in almost immediately ($t < 0.01$ μsec) and is accompanied by rapid decay of the plasma and the production of fast electrons. In the second case ($\nabla n \perp \mathbf{E}$) anomalously strong absorption sets in after a time delay ($t_{delay} = 1$-3 μsec) related to the time for the leading edge of the plasma to become deformed when the wave pressure exceeds the plasma pressure ($E_0^2/8\pi \gtrsim n_e T_e$).

At the present time considerable attention is being devoted to the study of the interaction of electromagnetic radiation with small volumes or bunches of plasma whose characteristic dimension l is smaller than or on the order of the wavelength of the radiation λ [1-10]. These studies will make it possible to account for the effects of such important features of real plasmas as their limited size and the inhomogeneous distribution of the charged particle density in them. It should be noted that extremely rapid changes in the density with a characteristic length $l \leq \lambda$ may develop even in plasmas which are initially extended in space and only weakly inhomogeneous ($l \gg \lambda$), for example in laser plasmas, because of the occurrence of various sorts of nonlinear processes in the neighborhood of a singular point with a critical density n_c ($\omega_{Le} \simeq \omega$). The study of the interaction of radiation with small plasmas is of great interest for a number of practical reasons as well as for its general physical significance, specifically, for rf heating of plasmas [1, 4, 7], for modeling phenomena in laser plasmas [10] and in the ionosphere [2, 5, 6], and so on.

There is special interest in research at high microwave powers where various processes which are no longer described by linear electrodynamics may occur. In particular, the collisionless absorption of electromagnetic radiation in plasmas becomes more important and is dominant when the total (electron–ion and electron–neutral) collision frequency ν is much less than the frequency of the electromagnetic field ω, i.e., $\nu/\omega \ll 1$. At the same time, many processes associated with this kind of absorption in various parts of the electromagnetic spectrum (optical, microwave, and rf) will have the same physical interpretation. This opens up extensive possibilities for modeling and generalization.

Experiments with small plasmas are typically done in the microwave region since waveguide technology makes it possible to measure directly the most important characteristics of the interaction of the radiation with the plasma, for example, the absorption or reflection coefficients of the incident radiation [6]. In addition, in the microwave region it is comparatively easy, as opposed, for example, to the case of the optical region [2-6], to obtain large values for one of the basic parameters which characterize the collisionless interaction (absorption), $v_E/v_{Te} \sim 1$-10 (where $v_E = eE_0/m\omega$ is the oscillatory velocity of the electron, $v_{Te} = \sqrt{kT_e/m}$ is its thermal velocity, and E_0 is the maximum amplitude of the field in the traveling wave).

At the present time two principal possibilities for the collisionless dissipation of electromagnetic energy in isotropic plasmas are known. The first absorption mechanism involves amplification of the field (a plasma resonance) in an inhomogeneous plasma near a singular point ($n = n_c$), which leads, at least in weak fields, to linear transformation of the incident electromagnetic wave into a Langmuir plasma wave at the same frequency. Here the incident wave must have an electric field component along the density gradient ($\nabla n \cdot \mathbf{E} \neq 0$). The plasma wave propagates toward the less dense plasma and undergoes collisionless Landau damping [2-5].

The second absorption mechanism involves nonlinear transformation of the electromagnetic wave into plasma waves (Langmuir and ion-acoustic) which appear as a parametric instability develops in the plasma. This process does not necessarily require the existence of a critical plasma density† and can occur with

†However, when $n \sim n_c$ the thresholds for these processes are minimal [6].

normal incidence of a wave on a plasma ($\nabla n \cdot \mathbf{E} = 0$). According to theory, the efficiency of both these transformation processes is determined to a great extent by the degree of inhomogeneity in the density n over the volume, by the maximum density n, by the way the density falls off at the edge of the plasma (smoothly or sharply on the scale of a Debye length $r_{De} \sim v_{Te}/\sqrt{4\pi ne^2/m}$), by the angle between ∇n and $\mathbf{E}$, and by the value of v_E/v_{Te}.

The collisionless absorption of microwaves in small volumes of isotropic plasma has been studied experimentally in a number of papers [2-4, 6-9, 11] over a wide range of $v_E/v_{Te} \sim 10^{-3}$–10 and of the plasma density $n \sim (0.1\text{–}5)n_c$. However, the experimental equipment and techniques have not always made it possible to accurately and uniquely identify the conditions which control effective absorption, especially in the case of high fields. In particular, it has been difficult to conclude which is the dominant absorption mechanism. The bulk of the work has employed gas discharge tubes in which the plasma has a relatively high electron–atom collision frequency $\nu_{e0}/\omega \sim 10^{-2}$–$10^{-1}$. Going to comparatively high microwave fields ($v_E/v_{Te} \sim 0.1$) may cause additional ionization of the gas and lead to a number of new effects (for example, relaxation-ionization oscillations) which complicate the isolation of simultaneous collisionless absorption processes. It should also be emphasized that there are in fact no reports in which the absorption coefficient has been measured directly in small plasmas with free boundaries, i.e., where the plasma is not in contact with the walls of a discharge tube or waveguide. We note that the presence of a free boundary plays a dominant role in collisionless damping due to linear wave transformation [2-5]. The efficiency of absorption has not been studied under conditions favorable to a plasma resonance. There is almost no information on the temporal evolution of absorption at large values of v_E/v_{Te} ($\gtrsim 0.1$). The absorption of radiation in a plasma may lead to decay of the plasma and to production of fast charged particles. However, appropriate studies for small plasmas have been reported in only a few papers [11, 12] for a rarefied plasma with $n < n_c$ when the plasma resonance is excluded. Furthermore, in a number of fundamental experiments on anomalous absorption often attributed to the development of a parametric instability, there is no information on the effect of the form of the plasma boundary or the density distribution and their possible time evolution on the efficiency of absorption as defined in terms of the absorption coefficient D^2. In addition, the techniques for measuring the absorption (or reflection) coefficient of electromagnetic radiation when the plasma "load" is rapidly changing during the microwave pulse were not adequately developed [12, 13].

In this paper, while taking all the above remarks into account, we propose to study the following basic problems which arise during the interaction of microwaves with small collisionless plasmas:

1. Experimentally examine the variation of the absorption coefficient D^2 of pulsed microwave radiation in the 10 cm band for a plasma with a free boundary over as wide as possible a range of electric field strengths E_0 of the incident wave ($v_E/v_{Te} = 10^{-5}$–5) under initial conditions which favor linear transformation or, in general, a plasma resonance. Compare experimental and computational (strictly speaking, determined reliably only for $v_E/v_{Te} \ll 1$) data. Study the temporal evolution of the absorption as well as the characteristics of the plasma as it decays under the action of the microwaves.
2. Examine the temporal evolution of the anomalous absorption in a narrow skin layer ($l \sim \lambda$ and $l \ll \lambda$) of collisionless plasma under initial conditions which simulate the normal incidence of a wave on a plasma layer and which, therefore, do not correspond to plasma resonance conditions. Study the dependence of the absorption coefficient D^2 on E_0 and the plasma density. Examine the temporal correlation between the appearance of fast charged particles and the development of anomalous absorption.

In order to make it easier to recognize the dominant mechanisms for collisionless absorption and the conditions for most effective absorption and aid in a comparison of the processes which accompany absorption, it was decided to conduct all these studies on a single waveguide apparatus and use a single type of diagnostic apparatus for all the experiments. A preliminary analysis was made of the conditions and methods for correctly measuring the reflection coefficient R^2 for microwave from a plasma, that is, from a load with rapidly changing properties [12, 13]. The possibilities of forming both a sharp leading edge on the plasma flow and a short burst of plasma with the aid of electric and magnetic fields were examined on an auxiliary apparatus.

1. Description of the Experimental Apparatus

These problems were investigated on the equipment that is illustrated schematically in Fig. 1.

The plasma is injected into a circular waveguide 11 (14-cm-diameter stainless steel) from a spark source 3 with a discharge over a Plexiglas surface. A movable short-circuiting piston 2 with an 8-cm-diameter aperture in its center is covered by a coarse copper grid. The plasma flow passes through this grid into the

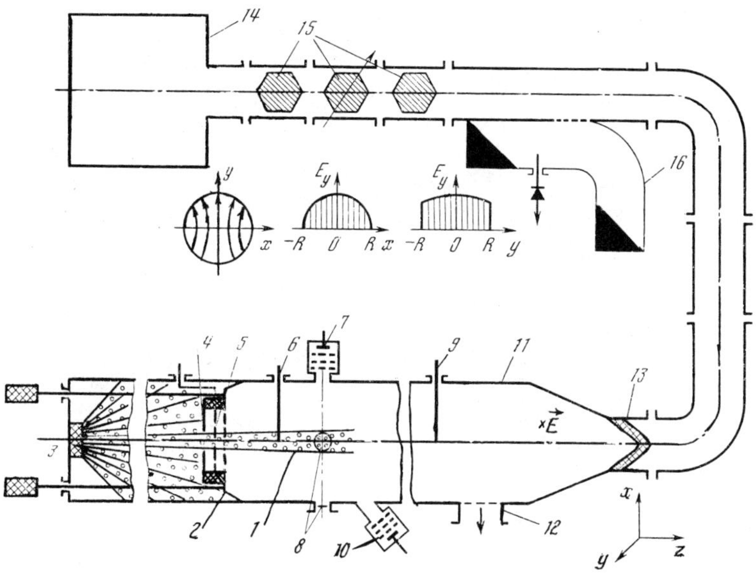

Fig. 1. A diagram of the apparatus and the field distribution in an H_{11} wave: (1) plasma flow; (2) movable short-circuiting piston; (3) plasma source; (4) diaphragm; (5) "blanking" grid; (6) screened single plasma probe; (7), (10) multigrid probes; (8) slab probe; (9) double probe; (11) waveguide; (12) pumping duct; (13) vacuum window; (14) microwave generator; (15) ferrite valves and attenuators; (16) directional coupler.

microwave interaction region. The speed of the front of the plasma flow was determined from the time of flight over a known distance from the source 3 and equaled $v_f \sim 10^7$ cm/sec. The electron temperature, measured with a double probe 9, was $T_e \approx 3$ eV.

An H_{11} wave with its electric field vector **E** perpendicular to the axis of the waveguide was propagated along the waveguide counter to the plasma. In these experiments the initial diameter of the flow is limited by means of a special diaphragm 4 with an aperture of diameter 3 cm $\ll \lambda$ located 2 cm in front of the grid on the piston. The radial density distribution is close to parabolic and is shown in Fig. 2 for distances of λ_g and $\lambda_g/2$ from the piston, where $\lambda_g \approx 20$ cm is the wavelength in the waveguide in the absence of a plasma. The density on the axis falls by at most a factor of 1.5 (due to thermal spreading) over a distance $\sim\lambda_g$.

It should be noted that the movable piston 2 was located a distance of 70-80 cm from the plasma source 3; thus, nearly fully ionized plasma with a total electron–ion and electron–neutral collision frequency $\nu_{coll}/\omega \sim 10^{-5}$ passed the piston. The frequency of electron collisions with the radial boundaries of the plasma was $\nu_T/\omega \sim 10^{-3}$. The pressure of the residual gas in the device was $p \sim 10^{-6}$ torr.

The plasma density n_0 at a previously chosen point on the axis of the axially symmetric flow was measured before the microwave pulse was turned on by means of a miniature screened single-electrode plasma

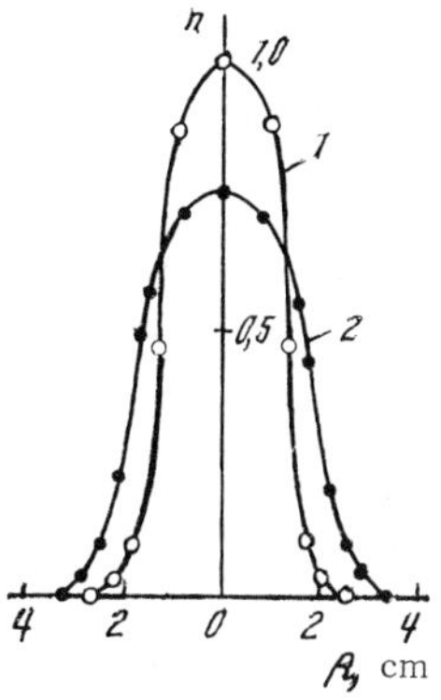

Fig. 2. The radial density distribution of the plasma at distances of $\lambda_g/2$ (curve 1) and λ_g (curve 2) from the piston 2.

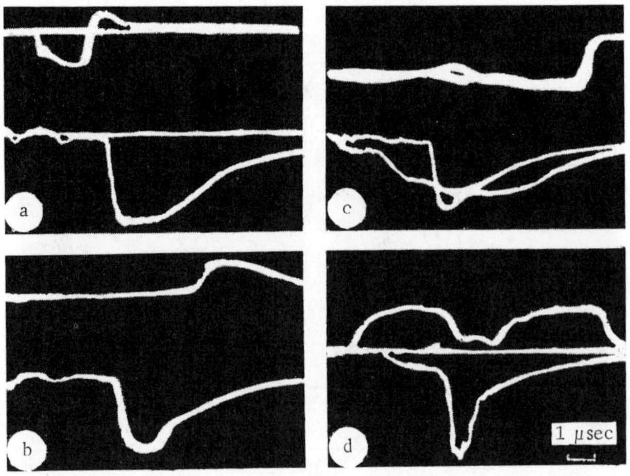

Fig. 3. Formation of plasma flows and bursts: (a) signals from the blocking generator (top) and probe 6 (bottom); (b) signals from the two plasma probes separated by a distance of 30 cm; (c) signals from the directional coupler of the reflected microwave pulse (top; leading edge not visible) and the probe 6 during formation of a steep front and in its absence (bottom); (d) signal from the directional coupler (top) and formation of a burst of length $\lambda_g/2$ (bottom).

probe 6 operating in the ion saturation current regime [14]. The outer diameter of the screen was 3 mm. In order to reduce the perturbations in the electromagnetic field of the H_{11} wave, the probe was fed through a side tube into the waveguide so that its axis was always perpendicular to **E**. The probe was calibrated absolutely by scaling from the known velocity and density of the ion saturation current at the collector. This calibration (for $n_0 = 2n_c$, where n_c is the critical density) was checked with the aid of an eight-millimeter interferometer.† Another check on the calibration (at $n_0 = 0.6n_c$) was made on the apparatus described in [8, 15] based on cutoff of the microwave signal at the working frequency. The results of the calibration checks agreed to within the measurement errors (not more than 15%) with the calculated (scaled) calibration values.

A grid with a mesh size of about 50 μm was located between the diaphragm 4 and the piston grid 2. When a negative voltage pulse (100 V, duration of 2-5 μsec) was applied to this grid with the aid of a blocking generator, a steep leading edge ($l_f \lesssim 3$ cm) was formed [12, 16] on the plasma flow. The operation of this shaping device is illustrated in the oscilloscope traces of Fig. 3a-c. Thus, before the microwave pulse was turned on it was possible to have a plasma flow of given length pass through the piston grid. We note that the length of the leading edge of the plasma flow could reach 50-60 cm if the shaping device were not turned on. (The apparatus for producing the plasma bursts is described in detail in Appendix 1.)

By establishing a fixed delay for triggering the blocking generator relative to the time the plasma source 3 was fired, it was possible in each series of measurements to stabilize the speed v_f. This is important for reading out the density from the probe (6) data. This was necessary as, even when the voltage on the source discharge condenser was constant ($U_{source} = 2$-5 kV, $C \approx 0.1\ \mu F$), the maximum density n_{max} and the speed of the plasma region with n_{max} could fluctuate within ±20% of the mean value from shot to shot.

If a constant negative cutoff voltage is applied to the grid 5 and at some time after the plasma arrives at the grid this voltage is compensated for a short time (about 1 μsec) by a positive pulse from the blocking generator, then it is possible to form a bounded plasma with a characteristic length $l \sim \lambda_g/2$ (on-axis density $n_{max} > n > n_{max}/2$). The trailing edge of this burst (see Fig. 3d, lower beam) is less steep than the leading edge [16]. In addition, to avoid breakdown and to increase the lifetime of the grid 5, the constant cutoff voltage applied to it was usually kept as small as possible which occasionally led to incomplete blocking and the formation of a small "step" in front of the leading edge of the plasma burst. This, however, had no effect on the nature of the interaction between the plasma and the microwaves.

†Here the piston 2 was removed and the central part of the waveguide, where the plasma probe 6 was located, was replaced by a glass tube of the same diameter as the waveguide. The radial plasma density distribution in the neighborhood of the probe 6 was practically uniform.

The currents and energies of the fast electrons and ions formed during the microwave–plasma interaction were measured by the retarding potential method [14] with the aid of the multigrid probes 7 and the slab probes 8 located in pipes on the sides of the waveguide. The plane probe was used for measuring the electron energies up to about 600 eV and consisted of a copper slab of diameter 3 cm covered with soot for reducing the effect of secondary electron emission (the secondary emission was less than 0.4). The inside surface of the waveguide near the measurement ducts was also covered with soot.

Microwave pulses were produced both with a low power standard signal generator and with a high-power magnetron generator 14. The two generators operated at the same frequency in the 10-cm band and by using regulating and decoupling ferrite valves 15 (total decoupling ~60 dB) the range of pulsed microwave power corresponded to a variation of E_0 from about 5×10^{-3} to 6×10^3 V/cm. The microwave absorption coefficient D^2 was defined as $D^2 = 1 - R^2$, where R^2 is the reflection coefficient. The reflected power was detected by a directional coupler 16 whose correct operation was ensured by suitable precision matching of the waveguide on both sides of the directional coupler [12, 13, 17]. When matching is not satisfactory, errors of up to 100% are possible. The criterion for good tuning of the waveguide circuit was the occurrence of minimal signal from the directional coupler when the short-circuiting piston 2 is moved. In our case, these oscillations did not exceed ±5% of the mean value. This determined the magnitude of the systematic error. (The problems in measuring the reflection coefficient are discussed in detail in Appendix 2.)

At small values of the field strength ($E_0 < 1$ V/cm) the signal from the directional coupler was detected by a P5-4B measurement receiver whose output was fed to an S8-2 recording oscilloscope. At high microwave powers a quadratic-characteristic crystalline microwave detector was used together with the same oscilloscope. The signal from the coupler was regulated with a D4-5 attenuator.

2. Measurements of the Absorption of Microwave Power and the Heating of Particles in Small Plasma Volumes with a Free Boundary

Absorption was studied for plasma flows and bursts of length λ_g and $\lambda_g/2$. The main purpose of the measurements was to compare the absorption coefficients in weak fields ($E_0 \sim 10^{-2}$ V/cm, $v_E/v_{Te} \sim 10^{-5}$) with those in strong fields (up to $v_E/v_{Te} \sim 5$) as well as to investigate the time-dependent dynamics of the absorption.

A typical oscilloscope trace of the absorption in weak fields for a plasma burst of length $\sim\lambda_g/2$ is shown in Fig. 3d, where the probe 6 is located a distance $\lambda_g/3$ from the piston. The appearance of absorption is very sensitive to the plasma density (see Fig. 3c). It was noted that absorption occurs in a plasma of length $\lambda_g/2$ only if the on-axis density n_0 at a distance of $\lambda_g/4$ from the piston (that is, at the field maximum of the standing wave) reaches or exceeds the critical density n_c (for $n_0 \leq n_c$ it is possible to neglect the variation in λ_g under our conditions). The threshold density n_c^* determined in this way from the onset of absorption agreed with the previously computed value of n_c to within an accuracy of up to ~10%.

This clearly makes it possible to relate the observed absorption to phenomena taking places at the singular point of the plasma [2-5], in this case to linear wave transformation, since the parameter characterizing the field amplification

$$\delta \sim (\omega L/v_E)^{2/3} \approx 30$$

(here L is on the order of the radius of the flow) which is much less than $\omega/\nu \sim 10^3$–10^5.

Figure 4 (smooth curve) shows a plot of the absorption curve in a plasma of length $\lambda_g/2$ as calculated using the results of [18]. There the absorption due to linear transformation of an electromagnetic wave traveling along a narrow cylindrical plasma column located on the axis of the waveguide with a parabolic radial density distribution was calculated. Since in our case a standing wave regime was realized, the calculation could be done only for the limiting cases of small and large absorption. In the first instance it is possible to neglect the change in the field caused by absorption, and the absorption efficiency per unit length of the plasma column can be expressed in terms of the damping constant given in [18]. In the second, we can neglect the wave reflected from the piston, and the results of [18] apply directly. Since the absorption must be a monotonically increasing function of the density over a wide range, it is possible to match the calculated parts of the absorption curve (dashed curve of Fig. 4). The same graph shows the experimental values of D^2 as a function of the density (in units of n_c^*).

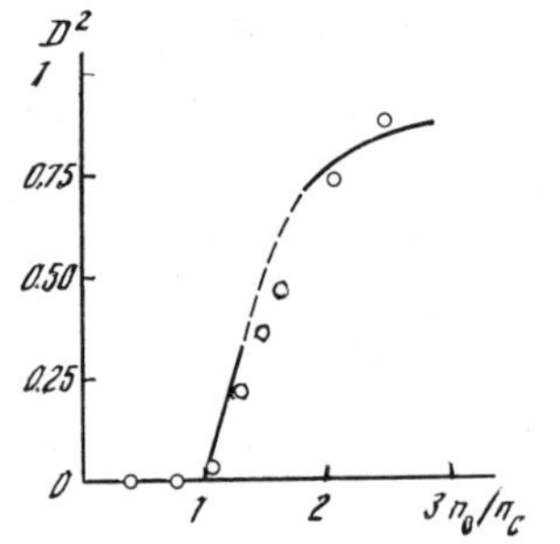

Fig. 4

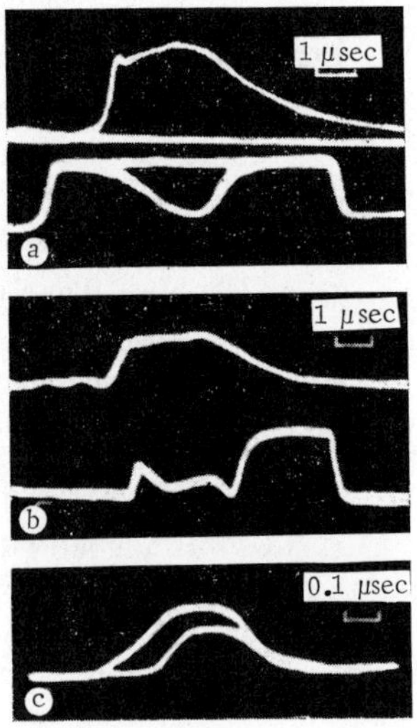

Fig. 5

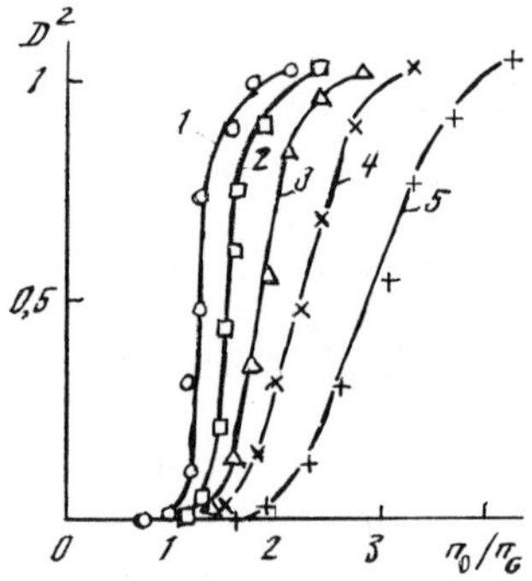

Fig. 6

Fig. 4. The dependence of the absorption coefficient in a plasma of length $\lambda_g/2$ on the density. n_0 is the on-axis density of the plasma burst in units of n_c^* for a distance of $\sim\lambda_g/4$ to the piston. The smooth curve shows the theoretical dependence; the points, experimental.

Fig. 5. Absorption in a plasma flow of length λ_g: (a) $E_0 \sim 0.01$ V/cm; (b) $E_0 \sim 50$ V/cm; the lower beam shows the microwave signal of the reflected wave and the upper, the signal from the probe 6; (c) $E_0 \sim 6$ kV/cm; the larger signal is the reflection from the piston 2 and the smaller is the reflection from a plasma with $n_0 \approx 2.5n_c^*$.

Fig. 6. The dependence of the absorption coefficient in a flow of length λ_g on the microwave power and initial density at a distance of $\lambda_g/4$ from the piston: (1) $E_0 = 0.01$ V/cm; (2) 100 V/cm; (3) 500 V/cm; (4) 1 kV/cm; (5) 6 kV/cm.

The dependence of the absorption coefficient on the power and initial plasma density was studied on a plasma flow of length λ_g. The risetime of the microwave pulse from the magnetron generator was ~0.2 μsec and the absorption was determined at a time† 0.3-0.4 μsec after the pulse was applied. The pulse, in turn, is turned on ~0.5 μsec after the steep leading edge of the plasma flow passes probe 6 which is located a distance $\lambda_g/2$ from the piston. A series of oscilloscope traces which illustrate the absorption process in a dense plasma is reproduced in Fig. 5. Figure 6 shows the variation in the absorption coefficient with E_0 and with the initial density (in units of n_c^*) at a point located a distance $\sim\lambda_g/4$ from the piston (calculated from the readings of probe 6). It is apparent that over a wide range of values of E_0 the absorption coefficient depends weakly on E_0 and is nonzero only when $n_0 \gtrsim n_c^* \approx n_c$. This statement is also true for the case $v_E/v_{Te} \gtrsim 1$ ($E_0 \gtrsim 1$ kV/cm), that is, when fast nonlinear (parametric) processes might be expected to develop [6]. The slight reduction in the absorption coefficient as the power rises for a given initial density is explained by the onset of expansion of the plasma and, therefore, by a reduction in the density even during the risetime of the microwave pulse since absorption at all power levels is inertialess. In this sense, the trace of Fig. 5c for $n_0/n_c^* \sim 2.5$ and $E_0 \sim 6$ kV/cm is very typical. Absorption occurs immediately from the time the microwave pulse is turned on, with $D^2 \sim 100\%$, which corresponds to the values of D^2 at a low power level. However, already before the end of the microwave pulse, that is, after 0.2-0.3 μsec, the absorption falls to 40%. It is this value which is plotted in Fig. 6. Therefore, at all microwave powers it is possible to reach $D^2 \sim 100\%$ if the initial density is sufficiently high. In all our experiments the absorption region within a cylindrical flow of length $l \sim \lambda_g$ that has a falling density along its axis is evidently limited to a size over which the on-axis plasma density $n_0 \geq n_c$.

Such efficient and inertialess absorption should and does lead to very rapid breakup of the plasma. For example, when $E_0 \gtrsim 200$ V/cm this time is $t \lesssim 0.5$ μsec if the density at the center of the flow $n_0 > n_c$. As an illustration we show oscilloscope traces (Fig. 7) of the ion current from probe 6 and the electron current to probe 7 for $E_0 \sim 1$ kV/cm, where probe 6 is located a distance of 8 cm from the movable piston. Figure 8 shows oscilloscope traces of the signals from the slab probes which reflect the appearance of fast electrons

†This is roughly 500-1000 periods T of the electromagnetic wave after the beginning of the pulse.

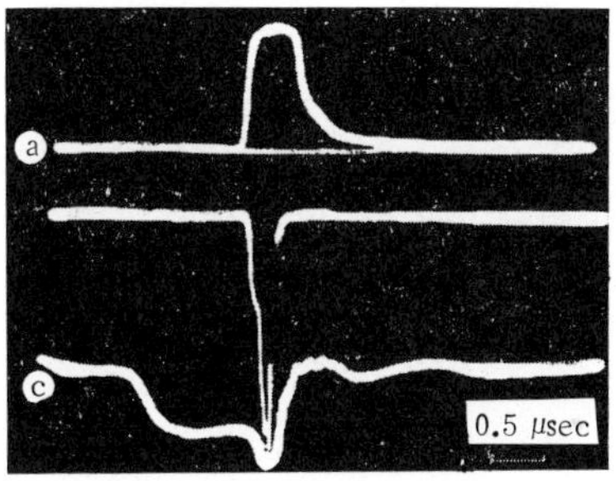

Fig. 7

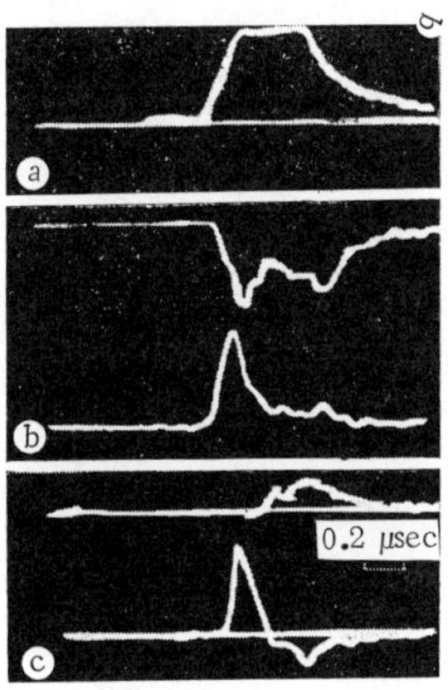

Fig. 8

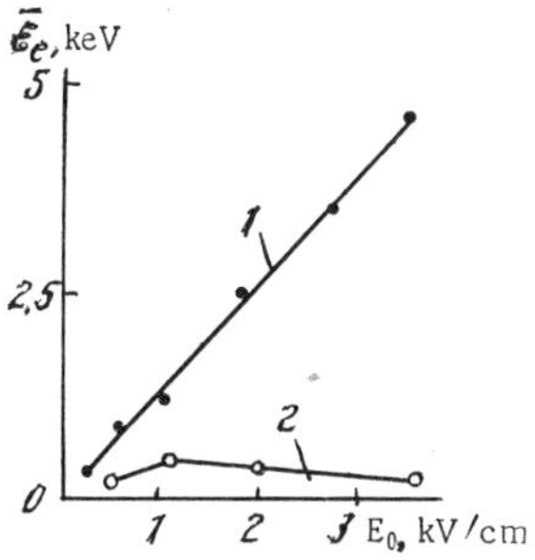

Fig. 9

Fig. 7. The decay of the plasma. (a) Microwave signal, $E_0 \sim 1$ kV/cm; (b) electron current signal on the multigrid probe 7; (c) ion saturation current on the plasma probe 6.

Fig. 8. Signals from the slab probes. (a) Microwave signal, $E_0 \sim 500$ V/cm; (b) electron current to the slab probes; top, current in a direction perpendicular to **E**; below, parallel to **E** with 10 times less sensitivity, $U_{probe} = 0$; (c) same as (b) with $U_{probe} = -600$ V.

Fig. 9. The dependence of the mean energy of the electrons $\bar{\mathscr{E}}_e$ on E_0 for $n_0 \approx 1.5 n_c^*$: (1) $v_e \parallel \mathbf{E}$; 2) $v_e \perp \mathbf{E}$.

and ions when the microwave radiation ($E_0 \sim 500$ V/cm) is acting on the plasma flow. Figures 9-11 show the values of the mean energy of the electrons in the expanding plasma and of their currents, as well as their distribution function as functions of the field strength E_0 in the incident wave, but for the same density $n_0/n_c \sim 1.5$ in the beginning of the flow (with a sharp front) as it passes the plane of probes 7 and 8 (the distance from piston 2 to the plane of the probes is $\lambda_g/3$). The ability to detect the fast electrons is determined only by the sensitivity of the apparatus and by the level of the thermal electron background.† In any case, in the range of E_0 from 5 V/cm to 4 kV/cm, the mean energy of the electrons $\bar{\mathscr{E}}_e$ (at a time of 0.3 μsec from the beginning of the pulse, corresponding roughly to the peak in the electron current to the probe) rises proportionally to E_0 (see Fig. 9). The numerical value of $\bar{\mathscr{E}}_e$ (in keV) is close to 1.5E (in kV/cm). The form of the distribution function of the fast electrons (along **E**) is also of interest and consists of two groups. As the power is increased the fraction in the high-energy portion of the distribution increases at the expense of the low-energy group. We note that the electron current along **E** is always roughly 10 times greater than that perpendicular to **E**. If the plasma density (on the axis of the flow) in the plane of probes 7 and 8 is below critical, for example $n_0 \sim 0.6 \times n_c$, then the currents are reduced by roughly a factor of 100 and the energy of the electrons does not exceed values determined by the rf potential [7, 19] (taking a slight increase in the field within the plasma into account). Thus, for $E_0 \sim 4$ kV/cm (corresponding to $v_E/v_{Te} \sim 4$) the maximum energy of the electrons is ~150 eV and the maximum energy of oscillation of an electron is greater than 200 eV, including the fact that the wave is standing (see also Appendix 3).

Roughly the same result was obtained for a rarefied plasma when the reflecting piston was replaced with an absorbing piston with a VSWR of 2.

Figure 12 shows the dependence of the maximum energy of the hot electrons on the on-axis density of the flow for a fixed field strength $E_0 \sim 500$ V/cm. For a density $n_0/n_c \sim 2$ the maximum energy of the fast electrons exceeds the oscillatory energy (in a vacuum field) by about 3×10^3 times and the thermal energy by 10^3 times.

The energy of the ion component of the expanding plasma, at least in its front, as determined from the delay time between the beginning of the microwave pulase and the arrival of the ions at probe 8, reaches several keV (for $E_0 \sim 500$ V/cm) and corresponds roughly to the maximum energy of the fast electrons (estimates for the hydrogen component).

†The proposed mechanism for heating of the electrons, which explains the practically inertialess appearance of fast electrons even in weak external fields under these conditions, has been discussed in [28].

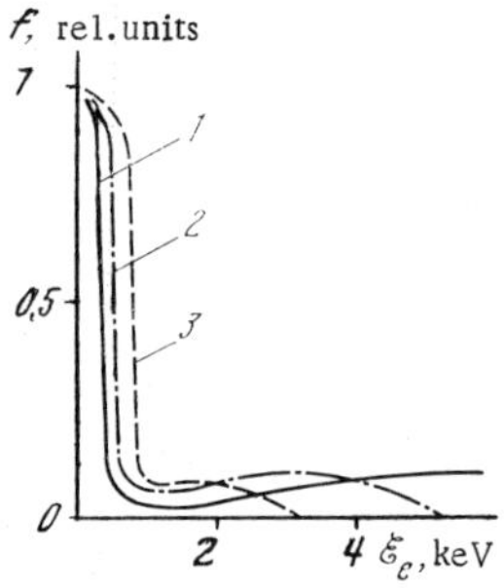

Fig. 10

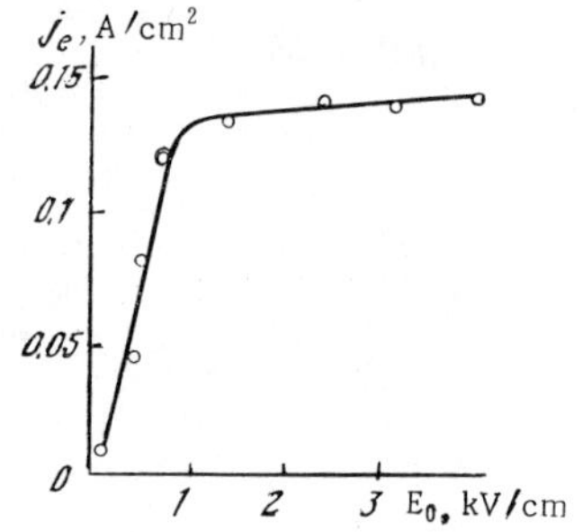

Fig. 11

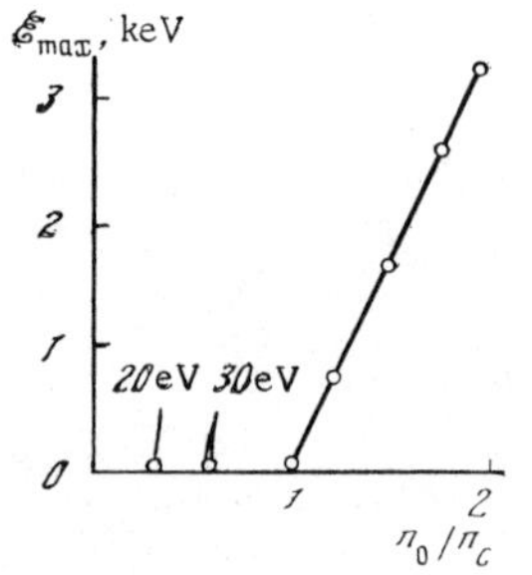

Fig. 12

Fig. 10. The electron energy distribution function for various E_0: (1) $E_0 = 4$ kV/cm; (2) 1 kV/cm; (3) 0.5 kV/cm.

Fig. 11. The dependence of the electron current on E_0 (along **E**).

Fig. 12. The dependence of the maximum energy of the fast electrons on the density at the axis of the flow for $E_0 \sim 500$ V/cm.

For moderate field strengths ($E_0 \sim 50$ V/cm) when the decay of the plasma during the microwave pulse is still relatively weak, a double probe at a "floating" potential was used to examine the possible change in the temperature of the bulk of the electrons in the front of the flow. Here a plasma flow with a steep leading edge "hit" the double probe 9 after the microwave pulse was over, as can be seen in the traces of Fig. 13. No change in the electron temperature (to within ±1 eV) was observed,† although such a flow emits fast electrons with energies of up to 200 eV during the interaction with the microwaves. For a uniform distribution of the absorbed microwave energy among all the plasma electrons we should expect an increase in the overall electron temperatures of $\Delta T_e \geq 10$ eV. One of the possible reasons for a lack of change in the temperature after the microwave pulse is over is that the absorbed energy is lost by intense heat exchange with the walls or with plasma lying beyond the piston and not subjected to the interaction with the radiation.

Therefore, the collisionless absorption coefficient of electromagnetic radiation in a small plasma with a free boundary is close to the value of the absorption coefficient in weak fields ($v_E/v_{Te} \sim 10^{-5}$), even in comparatively strong fields ($v_E/v_{Te} \sim 1$) and, at least, at the beginning of the electromagnetic pulse (over $t_1 \sim$ 500T-1000T). The absorption coefficient for weak fields, in turn, is determined by linear wave transformation.

Strong (for $n_0 > n_c$) absorption of electromagnetic radiation at all microwave field levels (up to $v_E/v_{Te} \sim 5$) remains practically inertialess, leads to rapid (over a time comparable to t_1) breakup of the plasma, and to a reduction or even the cessation of absorption. When $v_E/v_{Te} \gtrsim 1$ this effect becomes noticeable even in the front of the microwave pulse. During absorption fast electrons are generated with energies which rise in proportion to the field strength of the incident wave. The maximum energy of the ions in the expanding plasma is estimated to be on the order of the maximum energy of the electrons. For $n_0 < n_c$, neither noticeable absorption nor generation of fast particles (with energies exceeding the oscillatory energy of an electron) were observed.

The accumulated experimental data suggest that although the existing theory of linear transformation is valid, strictly speaking, only for weak fields, its quantitative conclusions may be applicable over a very much wider range. At least, the collisionless absorption in a small plasma volume may be extremely high (D^2 in tens of percent) even in strong electromagnetic fields, if the initial conditions are favorable for a plasma resonance ($n_0 > n_c$, $\nabla n \cdot \mathbf{E} \neq 0$).

3. Anomalous Absorption of Microwaves and Heating of Plasma Particles

An anomalously strong absorption of the electromagnetic radiation by the plasma was observed and first studied in a series of papers [8, 12, 13, 15, 20-22] devoted to the experimental study of the interaction of microwaves with collisionless isotropic plasmas inside waveguides. The effect of wall processes (secondary

†The electron temperature was measured in the leading, undecayed part of the plasma flow.

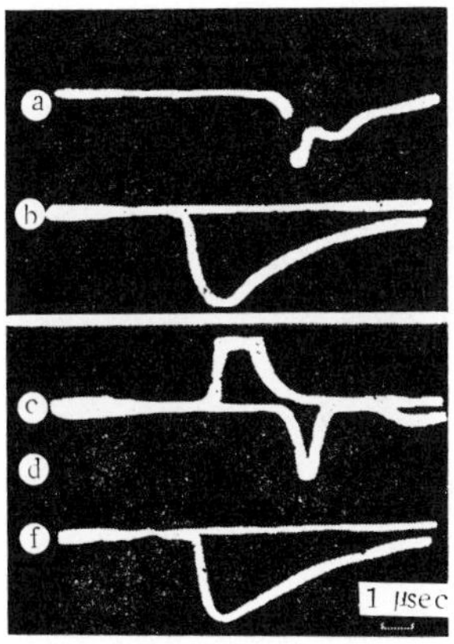

Fig. 13. Electron heating: (a), (b) signals from the double probe 9 and the plasma probe 6, respectively, in the absence of microwaves; (c) the microwave signal, $E_0 \sim 50$ V/cm; (d), (e) signals from probes 9 and 6 in the presence of microwave radiation. The electron temperature was measured to the leading, undecayed part of the plasma flow.

emission) was probably unimportant in at least a few of the experiments [13]. The absorption had a clearly nonlinear character and has traditionally been attributed to the development of a parametric instability [6]. However, the temporal evolution of the absorption remained unexplained, along with the possible role of "slow" nonlinear effects caused by the time variation of the plasma parameters (density, density distribution, temperature). These changes could be caused, for example, by the pressure of the rf field, by linear wave transformation phenomena that had not been completely eliminated [3], by the development of plasma instabilities, and so on. In principle, a similar type of phenomenon whose role increases in strong fields could cause a redistribution of the density such that absorption by means of linear transformation would again be significant, especially as in Sec. 2 we showed experimentally that highly efficient absorption is maintained, even in strong fields, when the initial conditions are favorable for linear transformation [23, 24]. In addition, it is necessary to determine more precisely the numerical values of the anomalous absorption coefficient and its dependence on the electric field strength E_0 of the incident wave since, at the time the first papers [8, 15, 20] were written, the techniques for measuring the absorption were not adequately developed.

This section is devoted to a study of processes in collisionless plasma under initial conditions which are not favorable for the linear transformation of waves, more precisely, for the nondecay transformation [24-27]. The experimental apparatus is similar to that described in Sec. 1. Thus, we limit ourselves here to a brief description of its characteristics and note the changes that have been made.

In this experiment a flow of collisionless plasma with $n_{max} > n_c$ that comes into contact with the walls propagates along a circular, 14-cm-diameter vacuum waveguide toward the microwave field. A movable piston 2 with an 8-cm-diameter aperture was usually located a distance of about 5 cm from the plasma source (Fig. 14). Because of thermal expansion the plasma flow becomes practically uniform over the cross section of the waveguide by a distance of about 50 cm from the piston. The characteristic dimension of the inhomogeneity in

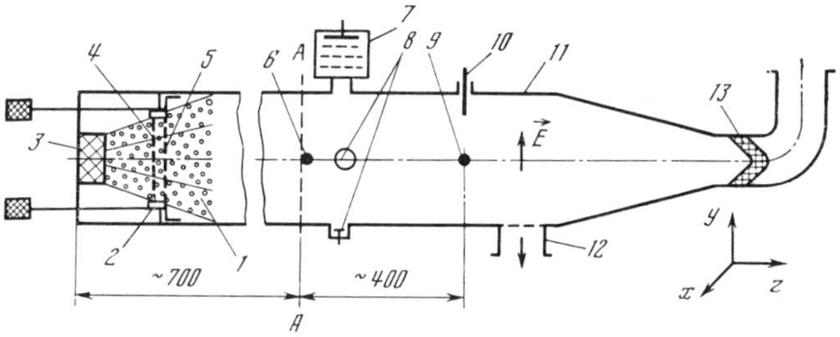

Fig. 14. A diagram of the apparatus: (1) plasma flow; (2) movable, short-circuiting piston; (3) plasma source; (4) grid for forming a steep front on the plasma flow; (5) grid in the aperture of the piston; (6) screened single-electrode plasma probe; (7) multigrid probe; (8) slab probe; (9), (10) screened single-electrode plasma probes; (11) stainless steel waveguide; (12) pumpout duct; (13) vacuum window; AA denotes the plane of the second position of the piston 2.

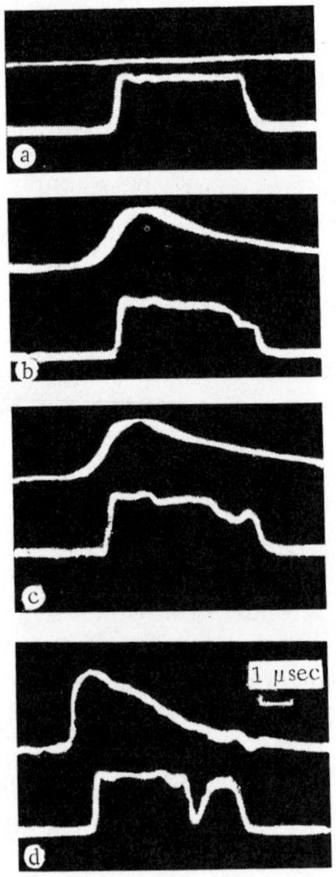

Fig. 15

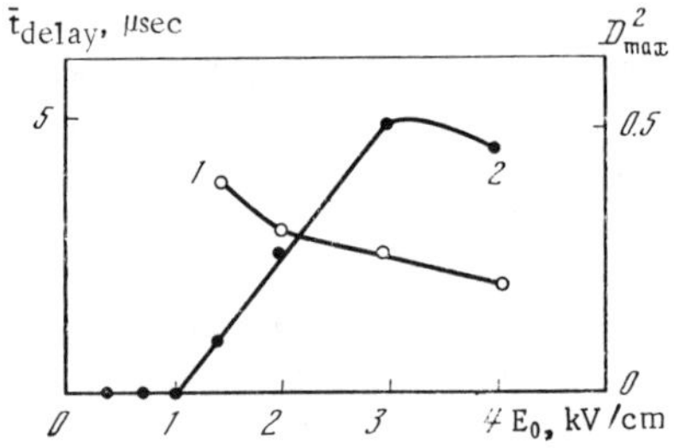

Fig. 16

Fig. 15. Absorption in a plasma flow: (a) reflected microwave signal without plasma; (b)-(d) signal with plasma for $E_0 \sim 2$, 3, and 5 kV/cm, respectively. The upper beam is the signal from the plasma probe 6; the lower, the signal from the directional coupler, proportional to the reflected microwave power. In (d) the plasma flow is formed with a steep front.

Fig. 16. The dependence on E_0 of the averaged time delays for the appearance of absorption (curve 1) and of the maximum absorption coefficient D^2_{max} (curve 2).

the plasma near the wall of the waveguide did not exceed 3 mm.† The large radial density gradient at the side of the plasma flow prevented intense linear transformation of the incident electromagnetic wave whose **E** vector is perpendicular to the axis of the waveguide (H_{11} wave).

It was also possible to form a steep leading edge on the plasma flow [16, 23] as it passed through a metal grid covering the aperture in the piston 2. In this case the piston was located about 70 cm from the plasma source and the diameter of the aperture was increased to 13 cm. The plasma density was monitored by the screened miniature probes 6 and 9 which were separated by a distance of ~40 cm. Probe 6 was located roughly 72 cm from the plasma source. The probes were calibrated absolutely and checked by the methods described in Sec. 1. The production of fast electrons was studied, as before, with the aid of the multigrid probes 7 and plane single-electrode probes 8 located in the side ducts of the waveguide. The duration and power of the 10-cm-band microwave pulse could be regulated and the maximum amplitude of the electric field strength of the traveling wave on the waveguide axis was $E_0 \sim 5$ kV/cm ($v_E/v_{Te} \sim 5$). The microwave power reflected from the plasma or from piston 2 was detected by a directional coupler with precision matching of the waveguide circuit on both sides of the directional coupler [17, 23, 24]. The error in the measurement of the reflection coefficient R^2 did not exceed ±5%. The pressure of the residual gas in the apparatus was $\sim 10^{-6}$ torr.

The basic measurements of the time-dependent efficiency and dynamics of the absorption were made when piston 2 was located near the plasma source 3 and the plasma was formed without a steep front.

†The upper bound for this dimension refers to a plasma located near the apertures for the side ducts with diameter ~3 cm. The plasma near the walls of the waveguide evidently has a boundary of thickness on the order of the Debye radius.

Figure 15 shows oscilloscope traces of the microwave power reflected from the plasma when the microwave generator was turned on at the time the plasma flow passes by probe 6 with a maximum density $n_{max} \sim 2n_c$. It is evident from these traces that significant absorption arises with a delay that decreases as E_0 is increased. The reproducibility of the reflected microwave pulses is poor; however, in some cases when $E_0 \gtrsim 2$ kV/cm, D^2 may reach roughly 50% for a short time. When $E_0 \lesssim 1$ kV/cm no significant absorption occurs over the entire pulse length $\tau = 4$ μsec.

Figure 16 shows the dependences of the time delay, t_{delay}, for the sure appearance of absorption ($D^2 \gtrsim 10\%$) and of the maximum value of D^2 on E_0. Both quantities are averaged over many pulses.

Under the experimental conditions, the wave ceases to propagate along the waveguide by the time the plasma density has reached $n \approx 0.6n_c$. This may lead to a large reduction in the electric field in the critical density region since the depth of the leading front of the flow $l_f \geq 2\lambda_g$, where $\lambda_g \approx 20$ cm is the wavelength in the waveguide. Thus an attempt was made to examine the effect on the absorption of steepening the front of the plasma. This was done by forming a steep plasma front so that $l_f \sim 3$ cm was $\ll \lambda_g$. This ensured that the condition $v_E/v_{Te} > 1$ was satisfied when $E_0 \sim 5$ kV/cm and in the critical density region, and yielded calculated values of the maximum growth rate for development of the parametric instability [6] of $\gamma_{max} \sim \omega(m/M)^{1/3} \sim 10^9$ sec^{-1} and of the effective collision frequency of $\nu_{eff} \sim \nu_{max}$.† Estimates for a homogeneous isotropic plasma show that this would correspond to the appearance of absorption at the level of a few tens of percent in the skin layer at the front of the flow almost immediately from the beginning of the microwave pulse.

However, in this experiment absorption was practically absent over the first 2 μsec at all initial densities from $0.3n_c$ to $\sim 2n_c$ and appeared with a delay $t_{delay} \geq 2$ μsec for $n > n_c$ (see Fig. 15d). In this case probe 6 was roughly 2 cm ahead of the piston and the microwave pulse was turned on when the length of the steep-edged flow which passed through the piston grid had reached $\sim\lambda_g/2$. The plasma, retaining its practically uniform distribution, had at that time already come into contact with the walls. The inhomogeneity in the density along the axis was less than 10%.

Strictly speaking, the measurements with the directional coupler only imply that the absorption coefficient D^2 at the initial time is less than 5%, the systematic error in the measurement of the reflection coefficient R^2. However, in this case it is probable that more accurate estimates of the absorption can be obtained from measurements of the currents and energies of the fast electrons which appear at the beginning of the interaction between the microwaves and the plasma. Thus, for $E_0 \sim 5$ kV/cm and $n \sim 1.5n_c$, in the experiments with a steep front the average electron energy was found to be $\bar{\mathcal{E}}_e \sim 0.2$ keV and the maximum energy was found to be $\mathcal{E}_e \sim 2$ keV.‡ The largest electron energies and current densities to the wall of the waveguide from the skin-layer at the front of the flow are obtained in the direction of the **E** vector. A corresponding estimate of the maximum energy transferred by electrons to the waveguide wall yields a value of the absorption coefficient over the first few microseconds of $D^2 \lesssim 0.5\%$. This estimate applies to all $E_0 \lesssim 5$ kV/cm.**

We note that for $n > n_c$ the fast electrons always appear almost simultaneously with the beginning of the microwave pulse. This statement is true all the way down to the weakest fields $E_0 \sim 100$ V/cm at which it is still possible to separate the fast electron current from the background current of thermal electrons from the plasma picked up by the slab probe 8. In these measurements the slab was located about 1 cm from the plasma boundary.

What happens to the plasma itself during the time t_{delay} prior to the onset of absorption? To answer this question, two plasma probes 9 and 10, located in the same plane at a distance of 40 cm from probe 6, were used. Probe 9 detected the plasma in the center of the waveguide and its axis was perpendicular to the **E** vector. Probe 10 detected the plasma at a distance of about 1 cm from the wall of the waveguide and its axis was parallel to the **E** vector.

Figure 17 shows oscilloscope traces of the signals from these probes which illustrate the deformation of the plasma front by the microwaves (the front was not shaped). There is a distinct delay in the propagation of the front due to the effect of the microwaves and there is a simultaneous rise in the density which is relatively less near the wall than in the center of the waveguide. For a pulse duration $\tau \sim 2.5$ μsec, when noticeable absorption occurs, intense decay of the plasma already apparently takes place. But, as before, its density is

†The inhomogeneity and nonstationarity of the plasma can lead to substantial increases in the threshold [6].
‡The measurements were made by the retarding potential method. Similar experimental $I(U_p)$ curves were obtained by averaging over many pulses and from measurements of $I(U_p)$ in a single pulse (see Appendix 3).
**For a rarefied plasma ($n_{max} < n_c$) the energy of the electrons did not exceed the oscillatory energy and the calculated value of D^2 was certainly less than 0.1%.

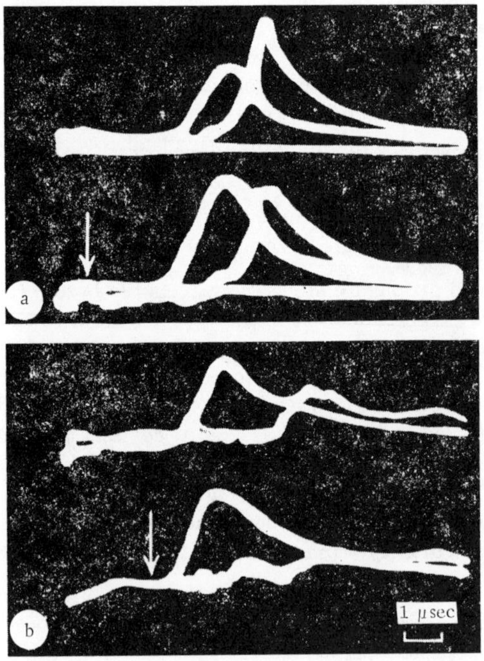

Fig. 17

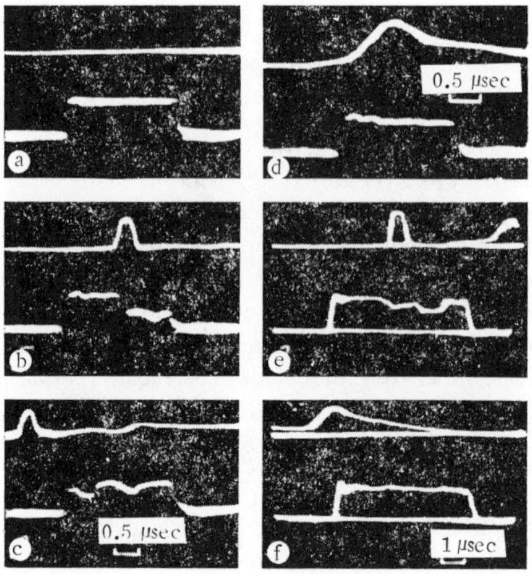

Fig. 18

Fig. 17. The deformation of the front of the plasma flow, $E_0 \sim 4$ kV/cm: (a) microwave pulse length $\tau \approx 1$ μsec; (b) $\tau \approx 2.5$ μsec. The upper beam is the signal from probe 9; the lower, from probe 10 (the sensitivities of the two are different). The signal on the left corresponds to plasma in the absence of the microwave field; that on the right is with the field turned on. The arrows correspond to the time of termination of the microwave pulses.

Fig. 18. The dependence of the reflected power on the time, relative to the firing of the gun, that the microwave generator is turned on: (a)-(d) $E_0 \sim 4$ kV/cm, scale 0.5 μsec; (e), (f) $E_0 \sim$ 0.1 kV/cm, scale 1 μsec. The upper beam corresponds to the signal from probe 6 and the lower, to the reflected microwave signal. In (b), (c), and (e) the signal on the left of the upper beam trace is "picked-up" from the gun.

relatively higher in the center of the waveguide than at the periphery. The delay in propagation and the relative rise in the density both decrease as E_0 and the duration of the microwave pulse are reduced while they remain almost unnoticeable for the threshold $E_0 \leq 0.7$ kV/cm, even when the microwave pulse length is $\tau \approx 5$ μsec, close to the actual lifetime of the plasma. This threshold value of E_0 corresponds roughly to equality between the wave pressure and that of the plasma in a reference frame coupled to the moving plasma flow, that is, to $E_0^2/8\pi \sim n_c T_e$.

Therefore, the experimental data suggest that the longitudinal pressure of the wave (which is greatest at the center of the waveguide), the transverse gradient of the rf quasipotential, and the residual linear transformation† at the sides of the plasma act together in the skin layer to cause a redistribution of the density with a maximum at the center of the waveguide. This increases the probability and efficiency of absorption by linear transformation.

The feasibility of this explanation for the observed phenomena is supported by oscilloscope traces (Fig. 18a-d) of the microwave power reflected from the plasma for different times of turning on the microwave power relative to the firing of the plasma source for $E_0 \sim 4$ kV/cm and $\tau \approx 2$ μsec. In this case the piston 2 is located near the source 3 and the aperture diameter is 8 cm. From the oscilloscope traces it is clear that absorption occurs at once, just as soon as the plasma, which is clearly inhomogeneous in its perpendicular cross section, passes through the piston grid into the microwave interaction region. Then, as the delay in turning on the microwave power is increased, the absorption is reduced and is practically absent at the beginning of the pulse if the delay reaches roughly 4 μsec. It is exactly then that the flow becomes almost uniform

†This can probably be related to the appearance of fast electrons and to the absence of a delay in their appearance, even in weak fields.

over the cross-sectional area of the waveguide. Qualitatively similar effects are observed even at significantly lower fields, for example, when $E_0 \sim 100$ V/cm (Fig. 18e-f).

It is also interesting to note that if a piston 2 with the same aperture diameter (8 cm) is located so far as 70 cm from the source, then noticeable absorption (up to 50%) already begins to occur during the second microsecond of the microwave pulse at fields of $E_0 \sim 200$-300 V/cm. However, at the beginning of the pulse for ~0.5 μsec D^2 does not exceed 10% even at 3 kV/cm (see [12, 13]). The increase in the absorption up to about 30% noted in these papers when the plasma completely fills the waveguide for a short microwave pulse ($\tau \sim 0.5$ μsec) corresponds to turning on the microwave pulse (see Fig. 18) at a time when the flow has evidently not become sufficiently uniform over its cross section.

To these remarks we must also add that including a weak constant multipole magnetic field $B \sim 100$ G ($eB/mc \ll \omega$, $B^2/8\pi > n_c T_e$) at the walls with a characteristic radial dimension of 1 cm led on the average over many pulses to an increase by almost a factor of 2† in the fraction of absorbed energy $W = \frac{1}{\tau}\int_0^\tau D^2 dt$ during a microwave pulse ($\tau = 4$ μsec at $E_0 \sim 3$ kV/cm). At the same time, control experiments with heating of the waveguide walls to 250°C (or coating them with soot‡) did not result in any noticeable changes in the absorption. A comparison of the qualitative and quantitative composition of the gas in the apparatus immediately after the plasma source was fired, with and without the microwave pulse, made with the aid of a time-of-flight mass spectrometer (MSKh-4) also revealed no significant differences.

These control experiments show that, although the absorption is very sensitive to the boundary conditions, it is apparently not due to any secondary emission processes at the waveguide walls.

The fact that absorption is so sensitive to the boundary conditions (at the side surface of the plasma flow), increases with the degree of inhomogeneity of the plasma over its cross section, and occurs after a relatively long delay probably indicates that the effect of the parametric instability on the absorption of electromagnetic radiation is small in our experiment. This is probably explained by the excessively rapid changes in the plasma parameters with time (both due to thermal expansion and to "slow" nonlinear processes). We note also that an investigation of the reflected microwave power using an S4-14 pulsed spectrum analyzer over the first two microseconds of the pulse did not reveal any significant deformations in the spectrum at all power levels (to a level of −20 dB from the peak at the fundamental), except for a shift in the frequency of the reflected signal by ~1 MHz to higher frequencies. This shift is fully explained by the Doppler effect caused by reflection of the wave from the plasma as it moves down the waveguide, and it occurs at all power levels.

It should also be pointed out that the conclusions in [8, 20] about possible anomalous absorption at fields $E_0^2/8\pi \ll n_c T_e$ ($\nabla n \perp \mathbf{E}$ for $E_0 \sim 200$ V/cm) in uniform (over its cross section) plasma flows in waveguides are not confirmed here and, apparently, were a result of the fact that an initially inhomogeneous (over its cross section) plasma was injected into a previously created microwave field.

To sum up, we may draw the following conclusions:

1. During the interaction of pulsed microwaves in a waveguide (H_{11} wave) with a homogeneous (over the cross section of the waveguide) column of collisionless plasma that has a density greater than critical and a leading edge of width $l_f < 0.1\lambda_g$-λ_g, noticeable absorption occurs with a delay $t_{delay} \sim 1$-4 μsec when $E_0^2/8\pi \gtrsim n_c T_e$. In this case the absorption coefficient may reach tens of percent.

At the very beginning of the microwave pulse (that is, when $t < t_{delay}$) and during an entire pulse (of duration $\tau = 4$-5 μsec, roughly equal to the lifetime of the plasma) with $E_0^2/8\pi < n_c T_e$ the absorption is less than the systematic error in the measurements, ~5%.

The absorption is very sensitive to the boundary conditions and increases with the extent of the initial inhomogeneity in the plasma over the cross section of the waveguide.

2. Absorption (for an initially homogeneous density distribution) occurs after the deformation of the leading edge of the plasma which causes formation of an inhomogeneous density distribution over the cross section with a maximum at the center of the waveguide. It may be assumed that the intense absorption which is then observed is caused by linear wave transformation effects whose efficiency increases as the inhomogeneity in the plasma column becomes greater.

---

†The random error in determining the average value of W is not more than 5%.

‡To reduce the possible influence of secondary emission processes.

3. Fast electrons appear almost simultaneously ($t \lesssim 0.01$ μsec) with the switching on of the microwave generator in the skin layer for $n > n_c$ at all power levels. This can apparently be explained by residual effects of linear transformation.

The coupling between "slow" nonlinear effects and the increased efficiency of dissipation of electromagnetic energy in plasmas which has been seen in this work can occur, in principle, in experiments on the anomalous interaction of microwave radiation with plasmas under free space conditions [9] where a number of "slow" nonlinear phenomena have been noticed.

We have studied the interaction of pulsed microwaves in the 10-cm band with collisionless plasma flows in waveguides, both under conditions which favor a plasma resonance and those which do not.

It has been shown that under favorable conditions for the plasma resonance, over a wide range of microwave powers ($v_E/v_{Te} = 10^{-5}$–5) the values of the absorption coefficient D^2 are close. Absorption sets in almost immediately ($t_{delay} < 0.01$ μsec) and is accompanied by rapid decay of the plasma and by generation of fast electrons.

Under initial conditions which impede the plasma resonance in a dense plasma ($n_{max} > n_c$, $\nabla n \perp \mathbf{E}$), anomalously strong absorption occurs with a time delay ($t_{delay} = 1$–3 μsec) when the wave pressure exceeds the plasma pressure ($E_0^2/8\pi \gtrsim n_c T_e$). Deformation of the leading edge of the plasma with predominant growth in the plasma density on the axis of the flow precedes the onset of noticeable absorption. This is a consequence of "slow" nonlinear phenomena which are caused by changes in the plasma parameters due to the action of the incident wave and which result in the creation of favorable conditions for processes associated with field amplification at a plasma resonance.

In rarefied plasmas ($n_{max} \leq n_c$) neither significant absorption ($D^2 < 0.1\%$) nor fast electrons (with energies above the oscillatory energy) were observed when $v_E/v_{Te} \leq 5$.

APPENDIX 1

Production of Short Plasma Bursts

In order to realize heating by means of radiation [1], fully ionized plasmas with dimensions smaller than the wavelength are required. In the 10-cm band the required plasma density is $n \sim 5 \cdot 10^9$–$5 \cdot 10^{11}$ cm^{-3}.

It has been proposed previously [29, 30] that compact bursts of plasma be formed by shortening plasma flows produced by pulsed spark injectors. The plasma was shortened by cutting off the rear portion of the plasma flow with pulsed transverse (relative to flow direction) magnetic fields ("magnetic gate") or longitudinal electric fields.

In order to reduce the effect of neutral atoms and obtain better natural grouping of the ions in velocity, it is usually necessary to position the shaping device at a flight distance of $l \geq 40$–50 cm from the injector. However, because of the large initial spread in the velocities of the ions, the leading edge of the plasma flow (velocity of mass motion $v \sim 10^7$ cm/sec, $n_{max} \approx 10^{11}$–10^{12} cm^{-3}) reaches several tens of centimeters in length so that the above-mentioned method does not make it possible to obtain plasma bursts of a few centimeters in length with densities $n \sim n_{max}$.

It was important to construct a device which could be used to form short plasma bursts from any part of the flow by simultaneously cutting off the front and back parts. Then the maximum density of the plasma in the burst is $n_{max} \sim 5 \cdot 10^{11}$ cm^{-3}. First it was necessary to solve the problem of forming a steep leading edge by creating a barrier for the front of the flow. The use of a pulsed magnetic field for this purpose is a complex technical problem involving the generation of a single current pulse (~10 kA) with a short fall time (~0.1 μsec). Up to this time the use of a longitudinal electric field has been limited by breakdown at plasma densities $n > 10^{10}$ cm^{-3}. The reason for this is probably that the separation of the plasma into its components and its subsequent disintegration [30] were a result of the creation of a potential barrier for the high-energy ions in the plasma flow ($W_i \sim 100$–200 eV) which required the application of a large ($U_{max} \sim 400$ V) positive pulse to a special cutoff grid.

In order to reduce the probability of breakdown and simultaneously increase the limiting plasma density, we decided to create a potential barrier for the electrons ($T_e \sim 5$ eV) by applying a negative voltage pulse to the cutoff grid. According to the theory of a highly negative probe located in a moving plasma [14], we should expect that the amplitude of the cutoff pulse need not exceed several tens of volts under our conditions if the grid size is ~30 μ.

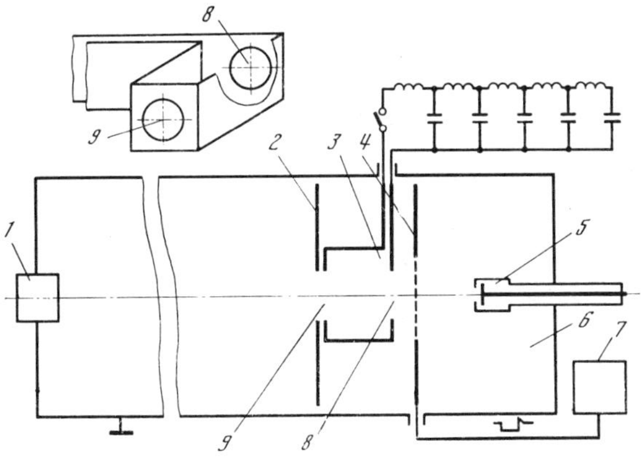

Fig. 1.1

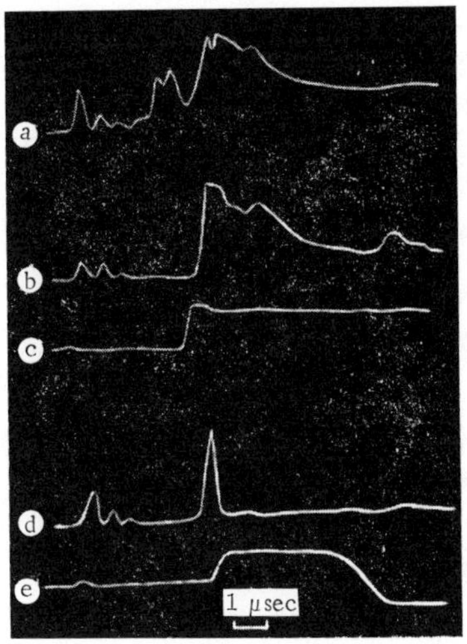

Fig. 1.2

Fig. 1.1. A diagram of the experimental apparatus: (1) spark-type plasma injector; (2), (4) dielectric and copper diaphragms; (3) magnetic gate; (5) screened single-electrode probe; (6) vacuum chamber; (7) blocking generator; (8), (9) holes.

Fig. 1.2. The production of a short burst from the plasma flow: (a) the signal from the collector of probe 5 in the absence of cutoff pulses; (b) same when a negative pulse is applied to the cutoff grid; (c) the negative pulse from the blocking generator; (d) the signal from the collector of probe 5 with two-sided shortening; (e) the magnetic field pulse.

Subsequent investigation showed that this method could indeed be used to cut off the front of the flow (risetime ≤ 0.3 μsec, density $n \sim 5 \cdot 10^{11}$ cm^{-3}, cutoff pulse amplitude less than 80 V); however, although it was possible, cutting off the back of the flow was ineffective since the trailing edge was very flat (lasting 1.5-2 μsec). Thus, in the following work a sharp trailing edge was produced by means of a "magnetic gate" [29].

Figure 1.1 is a diagram of the experimental apparatus used to study the two-sided shortening of plasma flows. The plasma was created by a spark source 4 with a discharge over a Plexiglas surface. The equipment included a dielectric diaphragm 2 with a 2-cm-diameter aperture, a "magnetic gate" 3, and a copper diaphragm 4 with a 3-cm-diameter aperture covered with a metal grid with a mesh size of roughly 0.03 mm (the cutoff grid). A screened single-electrode probe 5 was used for the measurements. The vacuum chamber 6 (10-cm-diameter stainless steel) was maintained at a pressure of $\sim 10^{-6}$ torr. The distance between the injector 1 and the diaphragm 4 could be varied from 30 to 60 cm and the distance between the probe 5 and the diaphragm 4 was approximately 2 cm.

Since the transparency of the cutoff grid was less than 60-70%, we might expect the occurrence of large ion currents to it (~1 A). Thus, the negative pulse to the grid came from a blocking generator 7 made with a dual tetrode GI-30 ($U_a \sim 2.5$ kV). The output voltage could be varied from 10 to 100 V and the pulse length was roughly 6 μsec.

The "magnetic gate," located a distance of 1.5 cm from diaphragm 4, was in the shape of a rectangular loop made of a 4-cm-wide copper strip. Two holes 8 and 9 of diameter 2 cm were made in opposite sides of the loop for passage of the plasma. The magnetic field (~1 kG) was produced by a discharge current from an artificial delay line ($U_L = 2$ kV, $I_L = 5$ kA). The magnetic field pulse was close to rectangular in shape (risetime ~0.3 μsec; duration ~5 μsec).

Some oscilloscope traces (Fig. 1.2) illustrating the "two-sided shortening" of the plasma flow include the signals from the collector of the screened probe 5 ($n_{max} \sim 5 \cdot 10^{11}$ cm^{-3}) with and without a negative pulse to the cutoff grid. The leading edge of the negative pulse from the blocking generator is not visible in the scan. Figure 1.2d shows the signal from the collector of the probe in the case of "two-sided" shortening. The probe detected the ion component of the plasma and operated in the saturation regime ($U_{probe} = -100$ V). On the left of the traces of the plasma signals one can see photoelectron current pulses from the probe collector caused by the luminosity of the injector when the discharge current is flowing.

Based on these and similar oscilloscope traces, we may conclude that this system actually makes it possible to form plasma bursts from any portion of the plasma flow. The minimum obtainable duration (full-width half maximum) of the plasma signal, ~0.3 μsec, corresponds to a burst of length ~3 cm with a transverse dimension of ~3 cm. The maximum density in the plasma bursts is $\sim 5 \cdot 10^{11}$ cm^{-3} for the maximum density in the plasma flow at the entrance of the "magnetic gate," $\sim 3 \cdot 10^{12}$ cm^{-3}.

APPENDIX 2

Measurement of the Coefficient of Reflection of High-Power Microwaves from a Nonstationary Load

In many physical experiments on the interaction of plasmas with high-power pulsed microwaves, one of the main problems is the measurement of the reflection coefficient [3, 6, 8, 13]. It is often important to obtain the time dependence of the reflection coefficient of the wave from the plasma in a single microwave pulse.

Direct measurements of the reflection coefficient can be made in practice only in waveguide structures where the plasma is inside an evacuated metal volume that is a part of or an extension of the waveguide circuit [3, 8, 13]. At high incident microwave powers ($P = 10^5$-10^6 W) the plasma properties (density, volume, effective collision frequency, etc.) may change sharply over a time t much shorter than the duration τ of the microwave pulse which is usually 0.5-3 μsec. In addition, the plasma itself as a whole may have a directed velocity relative to the waveguide (for example, if the plasma is created by means of plasma injectors). Thus, from the standpoint of microwave technique, the problem is to measure with a time resolution of 0.1 μsec the reflection coefficient of the microwaves from a nonstationary load in a single-pulse or slowly repeating pulse regime. Compared to the traditional microwave measurement techniques in a continuous regime with stationary loads [31], this imposes additional requirements whose neglect, as will be seen below, may lead to large absolute and relative (~100%) errors [12, 13, 17].

The high microwave power levels and the short duration of the processes being studied exclude the possibility of measurements with a movable microwave probe (measurement line). The techniques for measuring the voltage standing-wave ratio (VSWR) with the aid of several stationary microwave probes whose readings are then scaled to determine the reflection coefficient [32] become very time-consuming, both to adjust and to interpret, and, most importantly, they become inaccurate since for large reflectivities (~100%) the signal from the probe to the oscilloscope may change its amplitude by a factor of 100 during the pulse. Thus, under these conditions the only suitable method was the use of a directional coupler with weak coupling which, since it is inserted in series in the waveguide circuit, should react only to transmitted or reflected microwaves [3, 8, 13].

The directional coupler measurement device (Fig. 2.1a) consists of the main waveguide and an auxiliary waveguide parallel to it which are coupled weakly through the wide or narrow wall by means of a special system of slits or apertures [31, 33-37]. Two matched loads (I and II) are mounted on the ends of the auxiliary waveguide. A probe whose output goes to a quadratic detector is placed in front of one of the loads. Then, depending on the direction in which the coupler is connected, a signal proportional to either the input or reflected wave falls on the probe. Usually during a physical experiment it is sufficient to monitor just the magnitude of the reflected signal since a marker signal proportional to the incident power can easily be obtained with the aid of the same coupler in the reverse wave arm if a short-circuiting plate or piston (100% reflection) is introduced in place of the plasma load. Then the reflection coefficient from the plasma as a function of the power will be equal to the ratio of the amplitude of the reflected signal to that of the marker signal.

Despite the apparent simplicity of the measurements using the directional coupler, if certain conditions to be considered below are not met, large errors may result. The main sources of these errors can be conditionally divided into five groups:

(1) inadequate matching of load I in the auxiliary waveguide which absorbs a signal proportional to the direct wave;

(2) inadequate matching of the circuit beyond the coupler toward the microwave generator;

(3) inadequate matching of the circuit beyond the evacuated experimental volume toward the coupler;

(4) a shift in the frequency of the reflected wave (because of nonlinear effects) outside the tuned bandwidth of the coupler and waveguide circuit; and

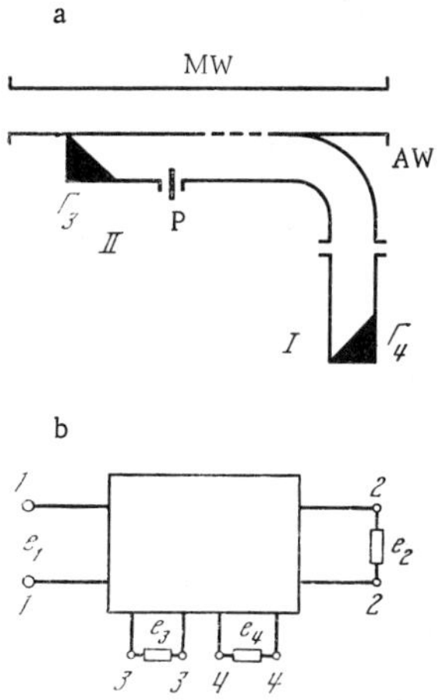

Fig. 2.1

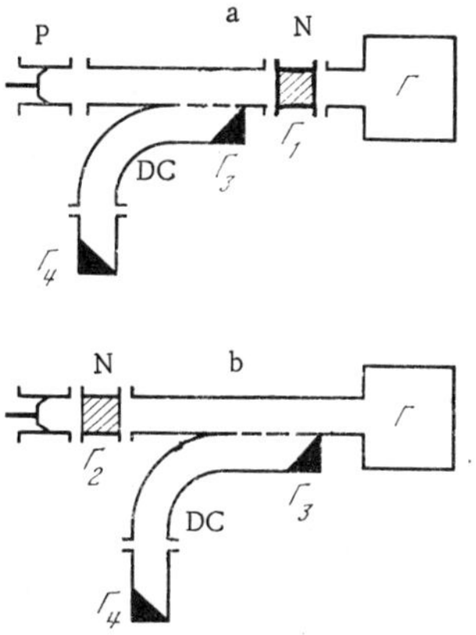

Fig. 2.2

Fig. 2.1. The directional coupler: (a) diagram of the directional coupler; MW, main waveguide; AW, auxiliary waveguide; P, probe; (b) equivalent circuit.

Fig. 2.2. The arrangements for inserting the directional coupler: P, movable short-circuiting piston; DC, directional coupler; N, inhomogeneity.

(5) a rotation of the plane of polarization (because of nonlinear effects) of the reflected wave relative to the incident wave which leads to multiple reflections from the plasma.

Here we shall of course neglect other possible, but trivial, errors such as deviations from quadratic detector characteristics, nonlinearities in the oscilloscopes, pickup, and the effect of changes in the load on the operation of the microwave generator.

From a physical standpoint the reason for the first three groups is the possible formation of resonances with a nonstationary load (although these resonances have relatively low Q) as a consequence of inhomogeneities in the waveguide circuit, in the experimental volume, or in the auxiliary waveguide of the directional coupler.

We shall now attempt to evaluate the magnitude of the error for each case separately and test the results experimentally.

1. In accordance with the equivalent electrical circuit, the directional coupler is an eight-pole network characterized by its scattering matrix [34-37]. However, in the case of measurement directional couplers, where the coupling between the main and auxiliary waveguides is weak (the intermediate attenuation is greater than 30 dB), the scattering matrix actually reduces to only two coefficients a_+ and a_- which characterize the directional properties of the coupling system: a_+ is the transmission coefficient in the "forward" direction and a_- is the transmission coefficient in the "backward" ("undesired") direction [36]. In the equivalent circuit (Fig. 2.1b) e_1 is the amplitude of the electric wave at terminals 1−1 passing in the main line from the microwave generator to the load, which is connected to terminals 2−2 (the transmission coefficient from terminals 1−1 to terminals 2−2 is practically unity) and e_3 and e_4 are the amplitudes of the electric fields at terminals on the arms 3−3 and 4−4 in the auxiliary line. We denote by $\Gamma_{22} = \Gamma_2 e^{i\varphi_2}$ the reflection coefficient (for the voltage) in the main line from terminals 2−2 and by $\Gamma_{33} = \Gamma_3 e^{i\varphi_3}$ and $\Gamma_{44} = \Gamma_4 e^{i\varphi_4}$, the reflection coefficients from the corresponding terminals of the auxiliary line.

For measurements with a single directional coupler (a coupler with a single arm) the (power) reflection coefficient Γ^2_{meas} to within second-order terms will be

$$|\Gamma_{meas}|^2 = e_3/e_3' = |\Gamma_{22} + \Gamma_{44} + a_-/a_+|^2, \tag{2.1}$$

where e_3' (the marker signal) corresponds to $|\Gamma_{22}| = 1$. Thus, in general, $|\Gamma_{meas}|^2$ differs from the actual value of the reflection coefficient $|\Gamma_{22}|^2 = \Gamma_2^2$. Usually for measurement directional couplers $a_-/a_+ \le 10^{-2}$ [the intrinsic directionality $D = 20 \log (a_+/a_-) \ge 40$ dB]. At the same time, for broadband loads it is extremely difficult to obtain a VSWR less than $\rho_4 \le 1.02$, which corresponds to $\Gamma_4 = (\rho_4 - 1)/(\rho_4 + 1) \ge 10^{-2}$. It thus follows that in most cases it is possible to neglect the error caused by the imperfection of the coupling system compared to the error caused by inadequate matching of the load at terminals 4–4. Then the maximum deviation in the measured value will be

$$|\Gamma_{meas}|^2 \approx |\Gamma_2 \pm \Gamma_4|^2 \simeq \Gamma_2^2 + 2\Gamma_2\Gamma_4. \tag{2.2}$$

Here the absolute value of the maximum possible error in the measurement may be (the uncertainty in the measurement of the true value)

$$\Delta|\Gamma_{meas}|^2 \approx 4\Gamma_2\Gamma_4,$$

while the relative error is

$$\delta = \Delta|\Gamma_{meas}|^2/\Gamma_2^2 \approx 4\Gamma_4/\Gamma_2.$$

For example, when a short-circuiting piston is moved in the main waveguide of the directional coupler (modulus $\Gamma_2 = 1$, phase φ_2 varying smoothly from 0 to 360°) the ratio of the maximum value of the measured quantity to the minimum will be

$$\left(\frac{A_{max}}{A_{min}}\right)^2 \approx \left(\frac{1+\Gamma_4}{1-\Gamma_4}\right)^2. \tag{2.3}$$

For an ideally matched load ($\Gamma_4 = 0$) this ratio is unity. However, if $\Gamma_4 = 0.1$ (corresponding to $\rho_4 = 1.2$), then $(A_{max}/A_{min})^2 \approx 1.4$ or the possible relative error may reach $\delta = 0.4$ or 40%. Therefore, the load at terminals 4–4 must satisfy very rigid requirements; specifically, its VSWR must be $\rho_4 \leq 1.02$. Only then will the measurement errors be less than 5% when $\Gamma_2 \simeq 1$.

2. In these discussions it has been assumed that the amplitude e_1 of the wave is independent of the reflectivity from the load that is being measured, $\Gamma_2 = \Gamma_{22}$. However, this is often not the case under actual conditions. In the following we shall examine an ideal directional coupler ($a_-/a_+ = 0$, $\Gamma_4 = 0$) when a short-circuiting movable piston is located in a lossless part of the main waveguide circuit and there is a small inhomogeneity in the circuit between the microwave generator and the directional coupler (Fig. 2.2a). This inhomogeneity can be formed by a ferrite rectifier, attenuator, switch, etc. Let the VSWR [31] of this aggregate inhomogeneity as measured from the side of the coupler be $\rho_1 \neq 1$. Then the (amplitude) reflection coefficient of the wave from the inhomogeneity may be written as

$$\Gamma_1 = (\rho_1 - 1)/(\rho_1 + 1) < 1.$$

Evidently, the wave reflection coefficient from an ideal piston is $\Gamma_L = 1$. The same will be true of the power reflection coefficient, $\Gamma_L^2 = 1$. In this case as the piston is moved along the waveguide the signal from the directional coupler should remain constant if the power from the microwave generator is held constant (by means of a sufficiently good ferrite decoupler). Actually, the signal does not remain constant. We shall, in fact, assume that a direct wave of unit amplitude passed through the inhomogeneity in the waveguide from the generator toward the piston. The amplitude of the return wave will also be equal to unity. However, this reverse wave is partially reflected from the inhomogeneity toward the piston with a reflection coefficient of Γ_1. This process will be repeated many times while the amplitude of the wave reflected from the inhomogeneity will fall by a factor of $1/\Gamma_1$ each time.

If the distance from the inhomogeneity to the piston (and back) is an even number of half wavelengths, then all the waves reflected from the inhomogeneity will be superposed in phase with the direct wave, and the total amplitude A of the wave propagating toward the piston (and, or course, from the piston) will be the sum of an infinitely decreasing geometric progression with a first term equal to 1 and a denominator of Γ_1; i.e.,

$$A_{max} = 1/(1 - \Gamma_1).$$

If, however, the distance from the inhomogeneity to the piston and back is an odd number of half wavelengths, then the waves reflected from the inhomogeneity are combined with the direct wave alternately in phase and out of phase so that the total amplitude of the wave propagating toward the piston (and back from it) is

$$A_{min} = 1/(1 + \Gamma_1).$$

Therefore, an ideal directional coupler which registers a signal that is proportional to the square of the amplitude (i.e., proportional to the power) will give different readings, depending on the position of the piston, whose maximum ratio is

$$\left(\frac{A_{max}}{A_{min}}\right)^2 = \left(\frac{1+\Gamma_1}{1-\Gamma_1}\right)^2. \tag{2.4}$$

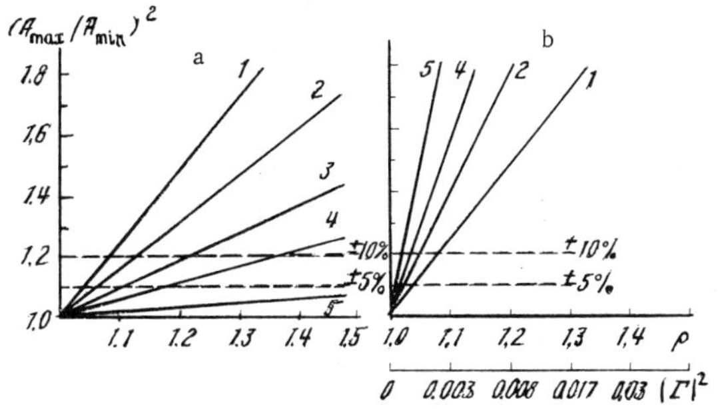

Fig. 2.3. The dependence of $(A_{max}/A_{min})^2$ on ρ for various Γ_L: (a) Eq. (2.5); (b) Eq. (2.6). (1) $\Gamma_L^2 = 1$; (2) $\Gamma_L^2 = 0.5$; (3) $\Gamma_L^2 = 0.25$; (4) $\Gamma_L^2 = 0.1$; (5) $\Gamma_L^2 = 0.01$. The dashed lines correspond to the measurement errors.

In this case an ideal reflecting movable piston served as one of the possible transient loads when the (power) reflection coefficient of the wave was kept constant and equal to unity while the phase of the reflected wave was varied. The resulting oscillations in the amplitude of the detected signal as a function of the marker signal level (which, as can easily be seen, could also be random) could be mistakenly interpreted as the appearance of substantial "absorption" or even "generation" of electromagnetic waves by the piston. It should be noted that during setting up of the waveguide tract usually only the required VSWR for normal operation of the generator, $\rho_g = 1.1$-1.3 at the input of the circuit, is checked [33]. In the presence of the ferrite rectifiers and attenuators which are used as decouplers, ρ_1 may be different from ρ_g and be unknown but, as a rule, it is excessively large.

We note yet another consequence of inadequate matching. When the attenuation of the transmitted power is changed with the aid of an attenuator, ρ_1 may vary due to changes in the matching. This in turn may result in a false dependence of the reflection coefficient on the incident microwave power.

If the movable piston has a reflection coefficient of $\Gamma_L < 1$, that is, it is a partially absorbing load with a power reflection coefficient Γ_L^2,† then the amplitude of the first wave reflected from the inhomogeneity will be $\Gamma_L\Gamma_1$, of the second, $(\Gamma_L\Gamma_1)^2$, of the third $(\Gamma_L\Gamma_1)^3$, and so on. Following the above line of discussion, we obtain this formula for the maximum ratio of the readings from the detector on the directional coupler as the piston is moved:

$$\left(\frac{A_{max}}{A_{min}}\right)^2 = \left(\frac{1+\Gamma_L\Gamma_1}{1-\Gamma_L\Gamma_1}\right)^2. \tag{2.5}$$

Since $\Gamma_L\Gamma_1 < \Gamma_1$, the relative scatter in the readings taken from an ideal directional coupler is now reduced compared to the case of a completely reflecting piston ($\Gamma_L = 1$) (Fig. 2.3a).

3. We now consider the third possible source of errors. Let there be a piston capable of moving along a segment of waveguide at the opposite end of which an inhomogeneity is attached. In a real waveguide circuit the role of this inhomogeneity could be played by a dielectric cone which separates the vacuum and air parts of the microwave circuit, different sorts of bends in the waveguide, switches, and so on. In addition, as shown in Fig. 2.2b, an ideal directional coupler and an ideally matched waveguide are installed on the microwave generator side.

Let a wave of unit amplitude propagate from the generator through the inhomogeneity toward the piston. If we assume for simplicity that $\Gamma_2 \ll 1$, then a wave of amplitude Γ_2 is reflected from the inhomogeneity. A reverse wave of amplitude Γ_L moves from the short-circuiting piston which moves past the inhomogeneity with almost the same amplitude toward the generator. Depending on the distance between the piston and the inhomogeneity, it combines with the wave Γ_L in two limiting cases (in or out of phase), that is, the total reverse wave is $A_{max} = \Gamma_L + \Gamma_2$ or $A_{min} = \Gamma_L - \Gamma_2$. Then the maximum ratio of the detector readings for the directional coupler as the piston is moved will be (see Fig. 2.3b)

$$\left(\frac{A_{max}}{A_{min}}\right)^2 = \left(\frac{\Gamma_L+\Gamma_2}{\Gamma_L-\Gamma_2}\right)^2. \tag{2.6}$$

†The (power) reflection coefficient is often denoted by $|R|^2 = |\Gamma_L|^2$ [8].

In this case the change in the readings from the directional coupler are an objective indication of absorption; however, this absorption occurs, not in the ideally reflecting piston, but in a resonator formed in the waveguide circuit between the inhomogeneity and the piston as the latter is moved. In a real experimental device, the occurrence of this effect as the reflecting boundary of the plasma moves along the waveguide may result in false conclusions about the absorption of electromagnetic radiation by the plasma. This inhomogeneity may yield especially large relative errors if the moving load has a reflection coefficient Γ_L which is much less than unity. This is explained by the fact that the amplitude of the wave Γ_2 is now comparable to the amplitude of the wave reflected from the load Γ_L.

In real waveguide circuits with a directional coupler, all three of these sources of error may evidently exist simultaneously.

The question of the magnitude of the total error is now considerably more complicated. If the load has a constant reflection coefficient Γ_L during the experiment and only changes the phase of the reflected wave (as, for example, in the case of a movable piston), then it is possible to use the following considerations to estimate the maximum scatter in the readings from the directional coupler.

The absolute magnitude of the signal from the directional coupler is always proportional to the power of the direct wave propagating in the main waveguide of the coupler, which may itself change from a minimum to a maximum for a fixed generator power (see Paragraph 2) by a factor of $(1 + \Gamma_{max}\Gamma_1)^2(1 - \Gamma_{max}\Gamma_1)^{-2}$, where Γ_1 is the coefficient of reflection from the inhomogeneity between the coupler and the generator, $\Gamma_{max} \simeq \Gamma_L + \Gamma_2$, Γ_2 is the coefficient of reflection from the inhomogeneity between the load (plasma, piston, and so on) and the coupler (it is assumed that $\Gamma_2 \ll 1$). However, even when the direct and reverse waves have constant powers, the detector signal may oscillate (see Paragraph 3) from a minimum to a maximum by a factor of $(\Gamma_{max} + \Gamma_4)^2(\Gamma_{min} - \Gamma_4)^{-2}$, where $\Gamma_{min} = |\Gamma_L - \Gamma_2|$ and Γ_4 is the coefficient of reflection from load I in the auxiliary waveguide of the directional coupler.

Therefore, we finally obtain

$$\left(\frac{A_{max}}{A_{min}}\right)^2 = \left[\frac{1+(\Gamma_L+\Gamma_2)\Gamma_1}{1-(\Gamma_L+\Gamma_2)\Gamma_1}\right]^2\left[\frac{(\Gamma_L+\Gamma_2)+\Gamma_4}{|\Gamma_L-\Gamma_2|-\Gamma_4}\right]^2. \tag{2.7}$$

4. We must also make the following comments on the accuracy of the measurements. When the microwaves interact nonlinearly with the plasma there may be a shift in the frequency of the reflected wave [6]. If this shift, as registered, for example, by an S4-14 pulsed spectrum analyzer, goes outside the band over which the previously mentioned inhomogeneities in the waveguide tract have been matched, then additional measurement errors may arise. In this case it is necessary to make a compromise: widen the tuning bandwidth but, generally, at the cost of reducing the accuracy of the measurements.

5. The wave reflected for the first time from the plasma may change its polarization inside a circular waveguide due to nonlinear effects and not return completely into a square waveguide. The presence of this phenomenon can be conveniently checked by introducing a counterbalanced microwave probe [38] across the circular waveguide through dielectric mounts as shown in Fig. 2.4. If the **E** vector of the wave is not strictly perpendicular to the metal tube-screen, then a wave can propagate on its outside surface and, as experiments show, strongly distort the probe readings. These distortions in the microwave field can be observed from the changes which occur in them when, for example, the tube is shorted to the waveguide at one of the outlet holes in the waveguide. In this way it is possible to observe a rotation in the plane of polarization of $\geq 5°$. When there is significant rotation of the polarization, the magnitude of the resulting error must be analyzed specifically for each experimental set-up. In particular, the surface wave may be suppressed when a microwave absorbing coating or load is placed on the metal tube.

Therefore, for measurements with an accuracy of $\pm 10\%$ of the (power) coefficient for single reflection of electromagnetic waves from the transient plasma it is necessary to achieve careful matching of the inhomogeneities discussed in Paragraphs 1-3 to a VSWR level of $\rho = 1.02$-1.03. The measurement of such a VSWR is at the limit of capability of standard measuring lines. Thus, they were suitable only for checking the tuning to within a VSWR of $\rho \sim 1.1$, while final matching of the inhomogeneities in the waveguide circuit was done in situ using as a criterion the absence or minimization of oscillations in the signal from the directional coupler probe as the short-circuiting piston was moved in the waveguide tract (preferably within the experimental volume). Here it is essential, of course, that the microwave generator be well decoupled so that the motion of the piston cannot cause an accompanying shift in the generator frequency or change in the output power.

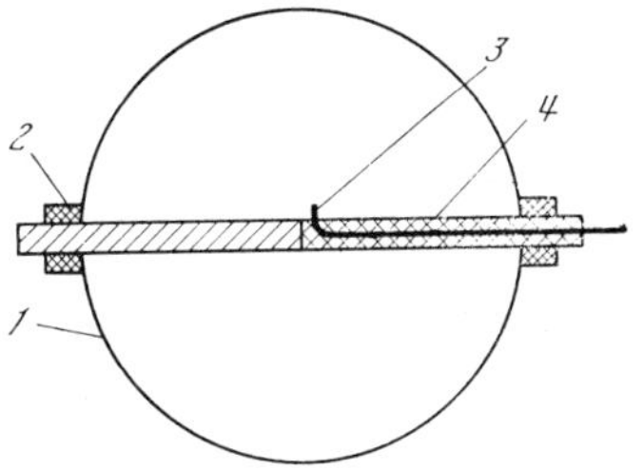

Fig. 2.4

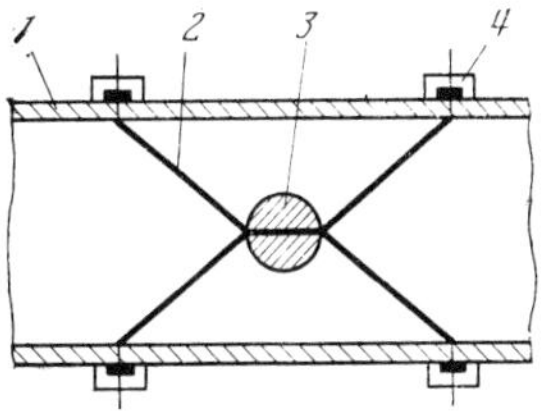

Fig. 2.5

Fig. 2.4. A sketch showing the introduction of a microwave probe with a symmetrizing counterbalance: (1) circular waveguide; (2) dielectric mount; (3) microwave probe antenna; (4) metal tube-screen.

Fig. 2.5. Illustrating tuning with the aid of a small sphere: (1) waveguide wall; (2) fiber; (3) metal sphere; (4) seal.

Experience showed that the use of movable metal pins (usually double- or triple-stub tuning transformers [31]) for matching the inhomogeneities at high microwave powers is undesirable since the impedance at the point of contact between the pin and the waveguide changes during the pulse. Thus even when the system is matched at the beginning of the microwave pulse, detuning may develop at the end (possibly due to microscopic breakdown). Fully satisfactory results were obtained with the aid of a contactless tuning method based on the movement of small metal (aluminum) spheres suspended on fine fibers (fishing line) inside the waveguide (Fig. 2.5). The fibers were brought out through rubber seals on the narrow walls of the waveguides so that the waveguide circuit could be pressurized. With a 20-mm-diameter sphere it was possible to match the inhomogeneities with a VSWR of 1.5 or less in a 10-cm-band waveguide. For convenience in tuning it was possible to use two spheres: a large one for "coarse" and a smaller one for "fine" tuning.

The typical arrangement of the elements of the waveguide circuit for pulsed measurements of the reflection coefficient from a transient load (plasma) is shown in Fig. 2.6. Each ferrite rectifier has a directionality of ~25-30 dB. The rectifier F_1 maintains the circuit matching when the power level is changed by the attenuator A (correspondingly for simultaneous changes in the attenuator VSWR). Similarly, in this case the rectifier F_2 keeps the circuit tuned during normal microwave operation. The total decoupling with the two rectifiers (~50-60 dB) is sufficient to eliminate any feedback from the changing load (plasma) to the microwave generator. We now present the results of a series of experiments intended to test the conclusions of Paragraphs 1-3.

The test stand included a directional coupler with coupling elements in the form of a system of slits [31, 34, 37] which ensured an effective directionality of the coupler of at least 35 dB over a wide range of frequencies. The load II in the auxiliary arm of the coupler had a VSWR of $\rho_3 \simeq 1.04$ and load I could be replaced or be tuned. In the first series of experiments (Fig. 2.7a) a waveguide section with a movable short-circuiting piston was located beyond the directional coupler. On the other side of the directional coupler in the waveguide circuit there was a pin with a variable insertion depth, an absorbing waveguide attenuator with damping of up

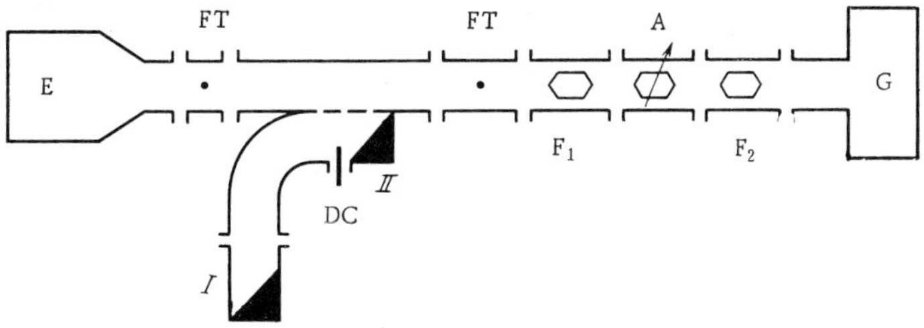

Fig. 2.6. A sketch of the experimental apparatus for measuring the reflection coefficient of a nonstationary plasma: E, experimental volume; DC, directional coupler; FT, fine tuning elements; F_1, F_2, ferrite rectifiers; A, attenuator; G, pulsed microwave generator.

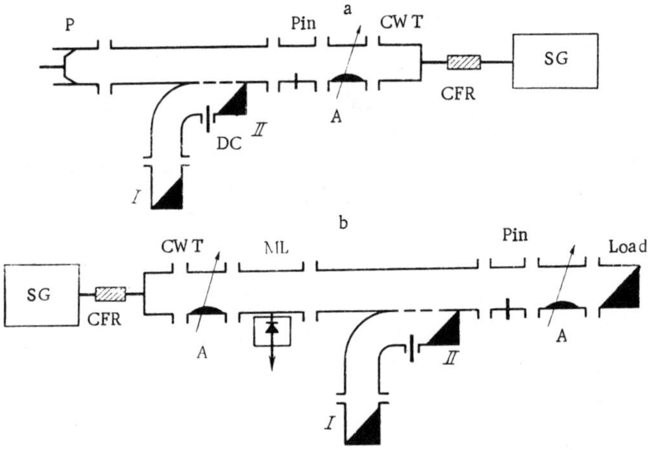

Fig. 2.7. A sketch of the measurements with the directional coupler: P, movable piston; DC, directional coupler; Pin, pin with variable insertion depth; A, absorbing waveguide attenuator; CWT, coax-to-waveguide adapter; CFR, coaxial ferrite rectifier; ML, measuring line; SG, standard signal generator.

to 30 dB, and a coax-to-waveguide adapter which was connected to a GSS-15 signal generator through a ferrite rectifier. The pin and the absorbing attenuator served as an aggregate inhomogeneity of the type examined in Paragraph 2, and the VSWR of this inhomogeneity was measured with the aid of a measuring line as shown in Fig. 2.7b. The signal from the auxiliary arm of the coupler corresponding to the power reflected from the movable piston was monitored with the aid of a square-law detector and a 28IM indicating measurement amplifier. The results of the measurements are shown in Table 1.

The data shown in the bottom line of the table correspond to additional in situ tuning done with the aid of a small sphere. These results have been confirmed at high power levels. The small difference between the computed and experimental values is probably due to a slight uncertainty in the VSWR of load I (see Fig. 2.1). The VSWR of load I, originally measured on the test stand, could change slightly after installation of the load on the directional coupler because of imperfections in the demountable connectors. In this regard the data on the VSWR of the pin and attenuator are more reliable since these measurements were made directly on the experimental device. Knowing the ratio $(A_{max}/A_{min})^2 = 1.7$ (third line of the table), which actually characterizes only the imperfection in load I, we may expect that the product of this ratio and the computed error caused by the mismatch on the signal generator side [see Eq. (2.4)] should coincide with the experimental values obtained for appropriate values of the VSWR of the pin and attenuator. In fact, for $\rho_1 = 1.58$, the "improved" computed value is $(A_{max}/A_{min})^2 = 3.9$, while the experimental value is $(A_{max}/A_{min})^2 = 4.3$. For $\rho_1 = 1.14$, this calculation yielded 2.0, while the experiment yielded 2.5. Thus, when the measurement error is taken into account, fairly good quantitative agreement is obtained between the experimental and calculated values.

Figure 2.8 shows the arrangement for the other experiment which was intended to test the effect of mismatches at the input to the experimental apparatus on the relative accuracy of the measurements of the absolute value of the coefficient of reflection from the movable load. Instead of a movable piston, a load with a variable reflection coefficient was placed in the waveguide section. The load in the directional coupler and the inhomogeneity on the generator side were carefully matched using small spheres as described previously.

TABLE 1

VSWR of load I, ρ_4	VSWR of pin with attenuator, ρ_1	$(A_{max}/A_{min})^2$	
		experiment	calculation with Eq. (2.7)
1.25	1.58	4.3	3.9
1.25	1.14	2.5	2.0
1.25	1.03	1.7	1.65
1.03	1.03	1.15	1.12
1.02	1.02	1.03	1.08

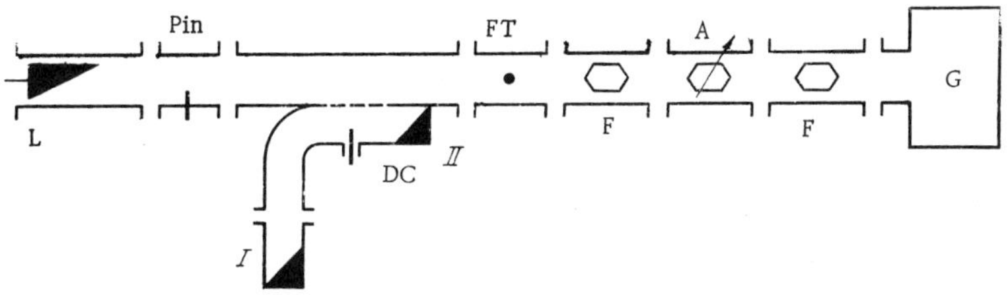

Fig. 2.8. A sketch of the measurement apparatus: L, movable load; pin, a pin for creating a mismatch at the input to the device. The other labels are as in Fig. 2.6.

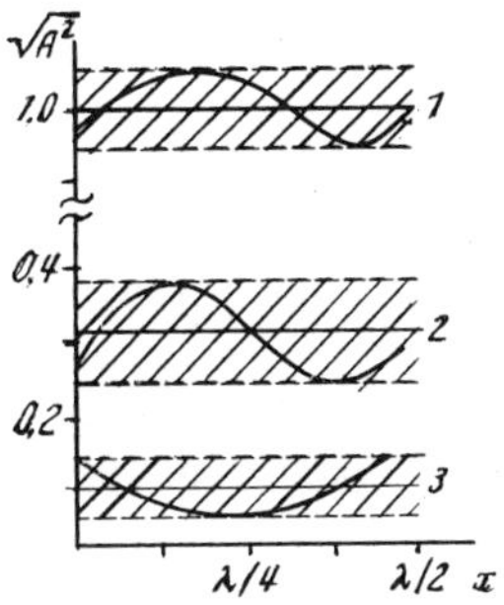

Fig. 2.9

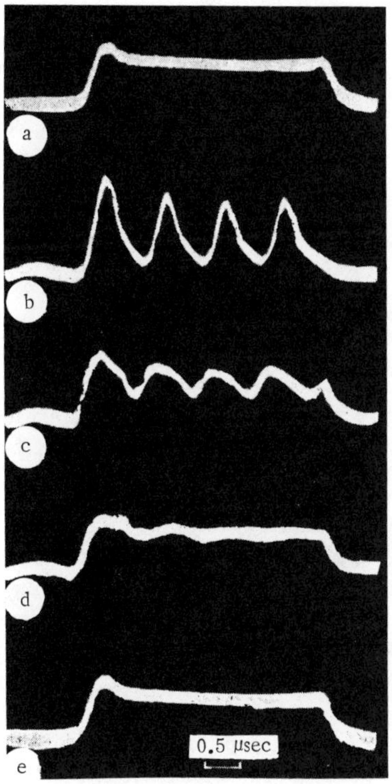

Fig. 2.10

Fig. 2.9. Dependence of the directional coupler probe readings on the location of the piston for various values of the VSWR of the movable load ρ_L: (1) $\rho_L = 100$; (2) $\rho_L = 2$; (3) $\rho_L = 1.22$.

Fig. 2.10. The reflection of microwave power from a moving plasma: (a) reflection from a short-circuiting piston with $\rho_1 = 1.02$; (b)-(e) reflection from a plasma piston with $\rho_1 = 3$; 1.4; 1.15; 1.02.

Figure 2.9 shows graphs illustrating the dependence of the readings from the directional coupler probe $\sqrt{A^2}$ on the location of the load as it is moved along the waveguide section. The VSWR of the mismatch at the input is $\rho_1 \approx 1.1$. It is apparent that as the reflection coefficient falls, the relative error in the measurement becomes larger. When there is a mismatch on the side of the apparatus with $\rho_1 \approx 1.4$, the fluctuations in the directional coupler probe readings as the fully reflecting piston is moved correspond to $(A_{max}/A_{min})^2 = 2$, which in turn corresponds to the theoretical estimates using Eq. (2.4). When $\rho_1 \approx 3$, $(A_{max}/A_{min})^2 \approx 9$, which indicates that the reflection from the piston can be substantially reduced at certain positions and, more precisely, that there is a resonant increase in the absorption in the metal waveguide section between the movable piston and the inhomogeneity at the input to the device. It is thus necessary to emphasize once again that large errors (up to 100%) are possible in measurements of the reflection coefficient of a nonstationary load using a directional coupler. These errors can be eliminated by careful matching (up to $\rho_1 \approx 1.01$) of the appropriate inhomogeneities in the waveguide circuit and also by monitoring the frequency spectrum and polarization of the reflected wave [12].

In order to conclude these investigations we made measurements on the experimental apparatus shown in Fig. 14 under conditions such that the leading edge of the plasma flow injected along the waveguide from the source played the role of the movable piston 2. In the experiments the electric field strength did not exceed $E_0 = 200$ V/cm and thus corresponded to the absence of microwave absorption in the moving plasma flow (see Fig. 16). The only uncompensated inhomogeneity which existed during the measurements was the inhomogeneity in the waveguide circuit beyond the directional coupler in the direction of the microwave generator. It was characterized by a variable ρ_1. The generator was turned on at the time when the leading edge of the plasma had moved a distance of 40 cm from the movable piston 2. The distance from the plasma source to the piston was 5 cm. Figure 2.10 shows oscilloscope traces of the microwave signal from the directional coupler. The reflected microwave power signals are shown for reflection from both the piston 2 and the front of a moving plasma for various degrees of mismatch in the waveguide circuit as characterized by the value of ρ_1. In the absence of any noticeable absorption (Fig. 2.10a and e, obtained for $\rho_1 = 1.02$), we have $(A_{max}/A_{min})^2 \approx 10$ for $\rho_1 \approx 3$. This ratio decreases monotonically with falling ρ_1. For $\rho_1 = 1.02$ $(A_{max}/A_{min})^2 \approx 1$. This is evidence of the virtual absence of power absorption in the microwave-reflecting metal or plasma pistons.

APPENDIX 3

Measurement of the Energy Spectrum of Fast Electrons Produced during the One-Time Interaction of Pulsed Microwaves with a Plasma

A characteristic effect of the interaction of microwaves with a plasma ($\omega_0 \lesssim \omega_{Le}$) is the generation of fast electrons [8, 21] whose energy is an order of magnitude or more greater than either the thermal energy of the plasma electrons or the maximum energy of oscillations of the electrons in the microwave field $W_{osc} = 2e^2E_0^2/m\omega^2$.

The energy distribution function of such fast electrons has been measured [8, 13, 22] at low pulse repetition rates of ~0.03 Hz. Graphs were plotted from the discrete points obtained by averaging the date from a large number of individual measurements. The resulting averaged graphs of the distribution function reflects reliably only the general integral properties of the phenomena being studied. The processes which occur in the plasma during the interaction with a powerful microwave pulse are not stationary in time. There is a certain statistical scatter in the plasma parameters during different pulses (n, T) and a discreteness in the energy measurements. Thus, the resulting electron energy distributions may significantly distort the details of the real energy spectrum, for example, indicating the presence of monoenergetic groups of electrons. In order to obtain more reliable and complete information on the energy spectra of the fast particles it is necessary to measure the distribution function over a wide energy range during a single interaction of a plasma with microwaves and over a time during which the parameters characterizing the interaction (plasma density n, microwave power P, current density of the fast electrons j) are practically constant.

Here we describe a method which was used to measure the entire spectrum of the fast electrons over a time of 100 nsec and covering a continuous energy range from 0.3 to 10 keV.

This technique was applied on a device (see Fig. 14) in which high-power type H_{11} microwaves ($v_E/v_{Te} \sim 1$) interacted with a cylindrical plasma in a circular waveguide. The spectrum of the fast electrons which left the plasma in the direction of the electric field vector **E** of the wave was measured. The energy of the electrons was analyzed using the retarding potential method with the aid of multigrid probes [39, 40] whose design and whose location in the experimental apparatus are shown in Figs. 3.1 and 3.2.

The analyzing voltage on multigrid probe No. 1 was a negative, linearly increasing (over a period of 100 nsec) pulse with an amplitude of from 2 to 10 keV delivered to one of the grids with a controlled delay relative to the time the fast electron current appears. For given parameters of the analyzing pulse, the electron current from the collector of multigrid probe No. 1 was recorded on the screen of an oscilloscope. This signal was a delay curve similar to the integral energy distribution function of the particles. In order that the measurements be accurate, it was extremely important to choose the duration and amplitude of the linear analyzing voltage correctly. The parameters of this pulse must satisfy at least the following two conditions: (a) the change ΔU_e in the electron current signal at the collector of the multigrid probe during the time the distribution function is measured must be small compared to the average value of that signal U_e over that time, that is,

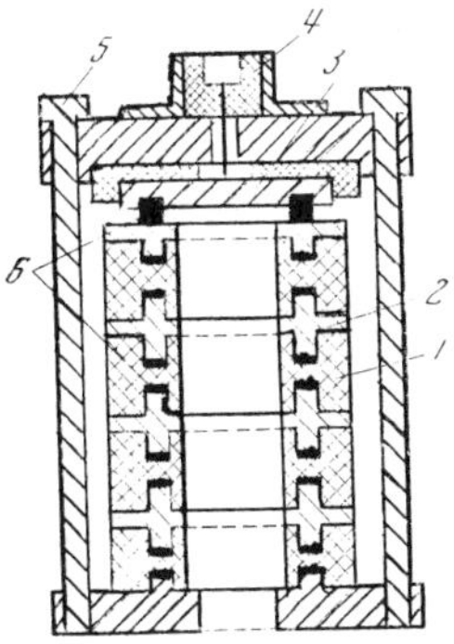

Fig. 3.1

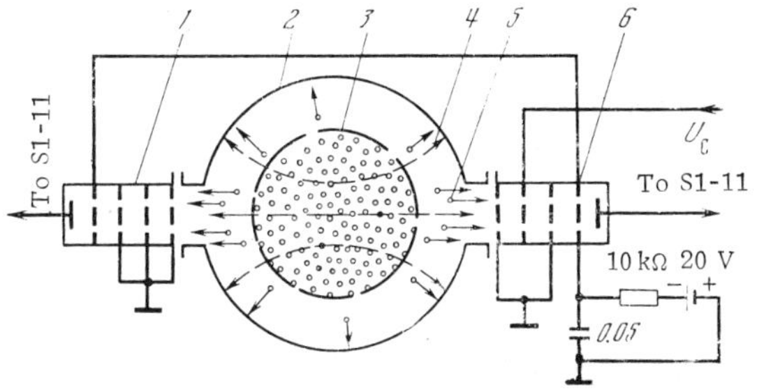

Fig. 3.2

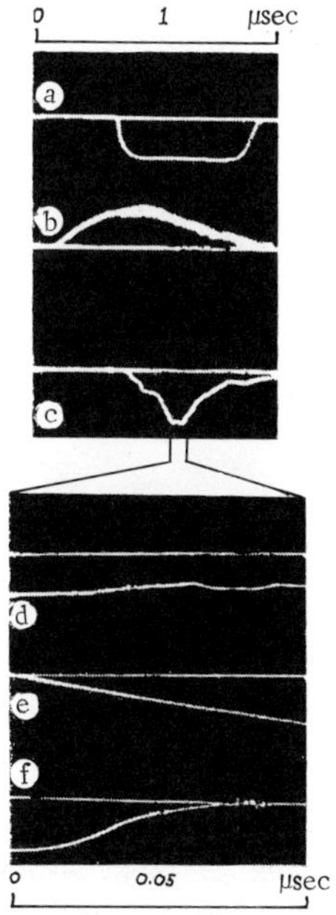

Fig. 3.3

Fig. 3.1. A cross-sectional diagram of the demountable high-voltage multigrid probe: (1) Teflon ring insulators; (2) metal ring grids; (3) collector; (4) collector lead; (5) pins; (6) rubber spacers.

Fig. 3.2. Illustrating the location of the multigrid probes on the apparatus (a cross section perpendicular to the axis of the waveguide): (1) multigrid probe No. 2; (2) 14-cm-diameter metal waveguide; (3) plasma shape; (4) electric field lines of the H_{11} wave; (5) direction of escape of the fast electrons; (6) multigrid probe No. 1.

Fig. 3.3. The signals at the S1-11 oscilloscope: (a) microwave pulse; (b) plasma ion current; (c) fast electron current pulse; (d) the straight portion of (c); (e) the linear analyzing voltage on the grid of multigrid probe No. 1; (f) the fast electron current pulse when the linear voltage is applied to the analyzer gap.

$\Delta U_e \ll U_e$, and (b) the change ΔU_g in the voltage on the analyzing grid during the time of flight across the analyzing gap of the multigrid probe No. 1, even for electrons with the minimum energy $W_{e\,min}$, must be small compared to the voltage U_g corresponding to this energy ($\Delta U_g \ll U_g = W_{e\,min}/e$). Condition (a) makes it possible to avoid errors connected with the variation in the magnitude of the fast electron current from the plasma during the measurement time, while condition (b) guarantees uniqueness in the electron energy determination since in this case, even for the electrons with the minimum energy, the electric field in the analyzer gap of the multigrid probe is practically constant during the time they cross the gap. In these measurements the analyzer pulse is a negative linear voltage which varies from 0 to 10 kV over a time $\tau = 100$ nsec.

In order to obtain the complete spectrum of the fast electrons during a single interaction between a plasma and a microwave pulse, a time interval was chosen over which the fast electron current to the collector of multigrid probe No. 1 remained practically constant. In our case this interval was generally 150-200 nsec. After this interval was chosen a linear voltage pulse was applied to the analyzer grid of multigrid probe No. 1. A current signal was recorded at the collector of multigrid probe No. 1 which in this case had the same form as the integral energy distribution function of the electrons (Fig. 3.3). The constancy of the flux of fast electrons during the measurement interval was monitored by the symmetrically located, completely identical, multigrid probe No. 2 to which an analyzing voltage was not applied. When there was no voltage on the grids of either probe, identical fast electron current pulses were recorded. The measurements were made on an apparatus (Fig. 14) for plasmas with densities $n > n_c$ and $v_E/v_{Te} \gtrsim 1$. An analysis of the oscilloscope traces of the integrated fast electron distribution function over energies of 0.3-10 keV showed that there were no steps on the traces; that is, there were no groups of fast electrons – provided the energy interval between them is at least 100 eV and the energy width of a group is at least 100 eV.

When there is a set of monoenergetic beams separated by small energy intervals, this apparatus (an S1-11 oscilloscope with a bandwidth of 250 MHz) is not able to reliably detect electron groups with such energy

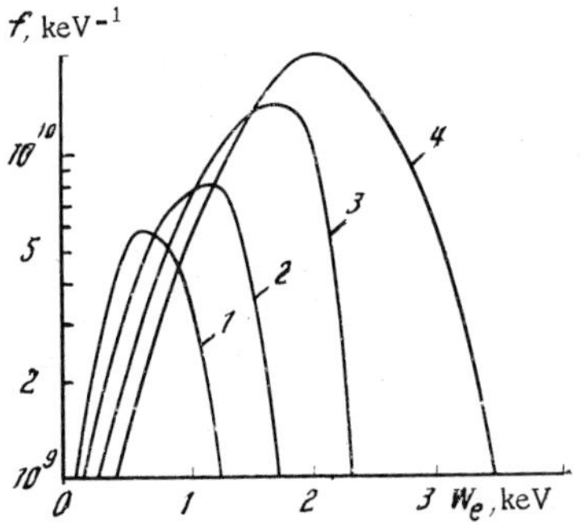

Fig. 3.4

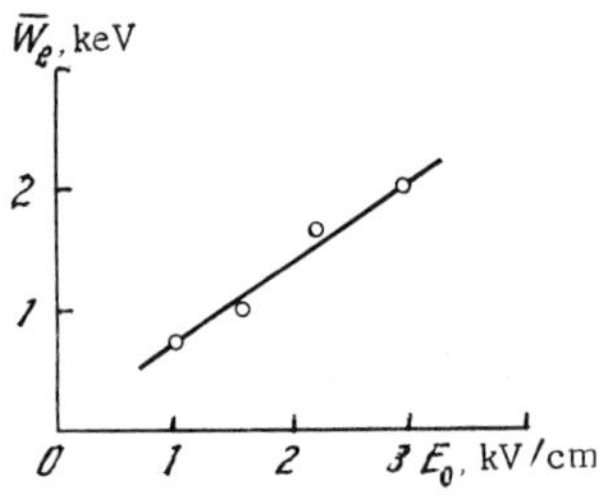

Fig. 3.5

Fig. 3.4. The energy distribution function of the fast electrons: (1) $E_0 = 1$ kV/cm; (2) 1.4 kV/cm; (3) 2 kV/cm; (4) 3 kV/cm.

Fig. 3.5. The dependence of the mean energy of the fast electrons $\overline{W}_e$ on E_0.

resolution. The relationship between the frequency bandwidth of the detection system and the energy resolution in this experiment is obvious. Thus, a group of electrons with energies of 5 ± 0.1 keV, which would produce a step of duration roughly 2 nsec on the delay curve, is practically unresolvable on the oscilloscope trace.

The delay curves obtained in a given pulse are in good agreement with those constructed by averaging over a large number of pulses under the same experimental conditions of plasma density n and field amplitude E_0. Figure 3.4 shows the differential energy distribution functions of the fast electrons for various values of the electric field amplitude E_0, obtained by differentiating the appropriate oscilloscope traces of the current (see Fig. 3.3f). In fact, $f_e(W) = \frac{dN}{dW} = \frac{1}{e}\frac{d}{dW}\int_0^{\tau} I\,dt' = \frac{\tau}{e}\frac{dI}{dW} = \frac{\tau}{ek}\frac{dI}{dt}$, where I(t) is the fast electron current obtained in the experiment; τ is the duration of the linear analyzing voltage on the probe; and k is the rate of rise of the analyzing voltage. The mean energy of the fast electrons, $\overline{W}_e$, increases almost linearly with E_0, the amplitude of the wave electric field. Figure 3.5 shows the dependence of the mean energy of the fast electrons on E_0.

LITERATURE CITED

1. V. I. Veksler, I. R. Gekker, É. Ya. Gol'ts, et al., At. Énerg., 18:44 (1965); Tr. Fiz. Inst. Akad. Nauk SSSR, 32:60 (1966).
2. P. E. Vandenplas, Electron Waves and Resonances in Bounded Plasmas, Wiley-Interscience, New York (1968).
3. V. E. Golant and A. D. Piliya, Usp. Fiz. Nauk, 104:413 (1971); Problems of Modern Physics [in Russian], Nauka, Leningrad (1974).
4. V. B. Gil'denburg, Candidate's Dissertation, NIRFI, Gorki (1965).
5. V. L. Ginzburg, Propagation of Electromagnetic Waves in Plasmas [in Russian], Fizmatgiz, Moscow (1960).
6. V. P. Silin, Parametric Interaction of High-Power Radiation with Plasmas [in Russian], Nauka, Moscow (1973); Usp. Fiz. Nauk, 108:625 (1972).
7. K. F. Sergeichev and I. R. Gekker, Tr. Fiz. Inst. Akad. Nauk SSSR, 73:3 (1973).
8. V. I. Barinov, I. R. Gekker, K. F. Sergeichev, O. V. Sizukhin, and V. E. Trofimov, Tr. Fiz. Inst. Akad. Nauk SSSR, 73:37 (1973).
9. G. M. Batanov and V. A. Silin, Tr. Fiz. Inst. Akad. Nauk SSSR, 73:87 (1973).
10. Lasers and the Thermonuclear Problem [in Russian], Atomizdat, Moscow (1973).
11. V. E. Golant, M. V. Krivosheev, and V. I. Fedorov, Zh. Tekh. Fiz., 40:382 (1970).
12. V. I. Barinov, I. R. Gekker, and V. A. Ivanov, Kratk. Soobshch. Fiz., No. 9, 7 (1973).
13. V. I. Barinov, I. R. Gekker, V. A. Ivanov, and D. M. Karfidov, XI Internat. Conf. on Phenomena in Ionized Gases, Prague (1973), Vol. 1, p. 329.
14. O. V. Kozlov, Electrical Probes in Plasmas [in Russian], Atomizdat, Moscow (1969).
15. K. F. Sergeichev and V. E. Trofimov, Pis'ma Zh. Éksp. Teor. Fiz., 13:236 (1971).
16. V. I. Barinov, Kratk. Soobshch. Fiz., No. 6, 8 (1971).

17. V. I. Barinov, I. R. Gekker, V. A. Ivanov, and D. M. Karfidov, Fiz. Inst. Akad. Nauk SSSR Preprint No. 53 (1974).
18. V. I. Fedorov, Zh. Tekh. Fiz., 41:680 (1971).
19. A. V. Gaponov and M. A. Miller, Zh. Éksp. Teor. Fiz., 34:242 (1958).
20. I. R. Gekker and O. V. Sizukhin, Pis'ma Zh. Éksp. Teor. Fiz., 9:403 (1969); IX Internat. Conf. on Phenomena in Ionized Gases, Bucharest (1969), p. 542.
21. V. I. Barinov, I. R. Gekker, O. V. Sizukhin, and É. G. Khachaturyan, Kratk. Soobshch. Fiz., No. 3, 41 (1971).
22. V. I. Barinov, I. R. Gekker, and V. A. Ivanov, Kratk. Soobshch. Fiz., No. 7, 8 (1973).
23. V. I. Barinov and D. M. Karfidov, Fiz. Plazmy, 1:638 (1975).
24. V. I. Barinov, I. R. Gekker, V. A. Ivanov, and D. M. Karfidov, VII Europ. Conf. on Controlled Fusion and Plasma Physics, Lausanne (1975), Vol. 1, p. 161.
25. V. I. Barinov, I. R. Gekker, V. A. Ivanov, and D. M. Karfidov, Fiz. Plazmy, 1:647 (1975).
26. R. Z. Sagdeev and V. D. Shapiro, Zh. Éksp. Teor. Fiz., 66:1651 (1974).
27. N. L. Tsintsadze and I. R. Gekker, VII Europ. Conf. on Controlled Fusion and Plasma Physics, Lausanne (1975), Vol. 2, p. 244.
28. V. I. Barinov, Fiz. Inst. Akad. Nauk SSSR Preprint No. 81 (1976).
29. É. Ya. Gol'ts and A. Z. Khodzhaev, Zh. Tekh. Fiz., 38:1960 (1968).
30. É. Ya. Gol'ts, Zh. Prikl. Mekh. Tekh. Fiz., No. 5, 113 (1966).
31. F. Tisher, Measurement Techniques for Microwave Frequencies [in Russian], Fizmatgiz, Moscow (1963).
32. I. R. Gekker and G. S. Luk'yanchikov, Zh. Tekh. Fiz., 35:1323 (1965).
33. I. V. Lebedev, Techniques and Apparatus for Microwave Frequencies [in Russian], Énergiya, Moscow (1964).
34. A. L. Fel'dshtein, L. R. Yavich, and V. P. Smirnov, Handbook of the Elements of Waveguide Technology [in Russian], Sov. Radio, Moscow (1967).
35. Measurement Techniques for Centimeter Waves [in Russian], Sov. Radio, Moscow (1949).
36. B. M. Mashkovtsev, L. Z. Benenson, and A. A. Khokhorev, Radiotekh. Élektron., Vol. 15, No. 4, 8 (1960).
37. J. Altman, Microwave Devices [Russian translation], Mir, Moscow (1968).
38. I. R. Gekker, in: Plasma Diagnostics [in Russian], Vol. 2, Atomizdat, Moscow (1968), p. 230; Fiz. Inst. Akad. Nauk SSSR Preprint No. 19 (1966).
39. V. A. Lavrovskii, Candidate's Dissertation, Institute of Radioengineering and Electronics, Academy of Sciences of the USSR, Moscow (1972).
40. V. A. Lavrovskii, E. G. Shustin, and I. F. Kharchenko, Pis'ma Zh. Éksp. Teor. Fiz., 15:84 (1972).

NONLINEAR EFFECTS IN THE PROPAGATION OF ELECTRON PLASMA WAVES IN AN INHOMOGENEOUS PLASMA LAYER

V. A. Silin

A review is given of some experiments on the dynamics of electron Langmuir oscillations and plasma waves in weakly inhomogeneous collisionless plasmas. Field amplification at a plasma resonance, collapse, and nonlinearities in the collisionless damping of waves are discussed. Experimental data on the deformation of the field distribution in the neighborhood of a plasma resonance are presented. It is established that, at high field amplitudes, in an inhomogeneous layer plasma waves fill the region from the critical density point n_c to $\sim {}^1/_4 n_c$.

INTRODUCTION

When an electromagnetic wave is incident on an inhomogeneous plasma layer with a peak electron density greater than critical, the most important processes occur in the region near the plasma resonance. If the angle of incidence of the wave is nonzero, there is a rapid growth in the field at the point $\varepsilon = 0$ (Langmuir oscillations). These oscillations are screened by the dense plasma and propagate into the less dense plasma in the form of electron plasma (Langmuir) waves in accordance with the dispersion relation $\omega^2 = \omega_{Le}^2 + 3k^2v_{Te}^2$.

Langmuir waves may be excited by the so-called linear transformation of transverse waves into longitudinal waves (see [1, 2]). In that case the necessary condition is the existence in the plasma of a region where the frequencies and wave vectors of the transverse and longitudinal waves coincide, i.e., $\omega_\perp = \omega_\parallel$ and $\mathbf{k}_\perp \simeq \mathbf{k}_\parallel$. As Piliya has shown [3], the transformation coefficient for the energy flux, which characterizes the efficiency of this process, may reach 0.4 for an optimum angle of incidence and for small wave amplitudes.

Langmuir waves can also appear in a plasma as a consequence of the development of parametric instabilities (see [4, 5]). Their excitation may have a nonlinear and nonlocal character [6].

The study of Langmuir waves is of particular interest because they can undergo collisionless damping in a plasma, which may be important in examining the prospects for heating plasmas by electromagnetic radiation.

If the amplitude of the fields in a plasma, including the fields of Langmuir waves, even approaches the order of magnitude of the characteristic plasma field $E_p = (3\varkappa T_e m\omega^2/e^2)^{1/2}$, then the behavior of the system becomes nonlinear. We shall first point out the change in the nature of collisionless damping of the waves, which decreases because of capture of particles by the wave and because of distortions in the distribution function [7].

The field of the Langmuir waves may be redistributed because of the modulational instability [8, 9] which leads to formation of narrow packets. The mutual actions of a strong wave on the medium and of the medium on the wave may lead to self-focusing or plasmon collapse [10]. Then regions of greatly reduced density, or cavities, must develop in the plasma. There is an increase in the amplitude of the field in such a region in accordance with the density change.

Experimental results from a study of the dynamics of the field and density in the resonance region of an inhomogeneous plasma [11, 12] show that these processes occur in a weakly collisional plasma, or, in any event, under conditions where the quasistatic one-dimensional problem is being studied and the initial field structure is determined by the metal structures used to deliver the field to the plasma.

The purpose of this paper is to review the experimental results obtained from a study of the field distribution of Langmuir waves in an inhomogeneous layer of collisionless ($\nu_{ei}/\omega_{Le} = 10^{-5}$) plasma [13, 14] and to analyze the possible effect of the nonlinearities associated with the Langmuir waves on the dissipation of microwaves in the plasma [15-17] and on the nonlinear transparency of layers of overdense plasma [17-19].

We note that in our experiment we have mainly studied fields and waves under conditions such that the linear transformation efficiency is rather high.

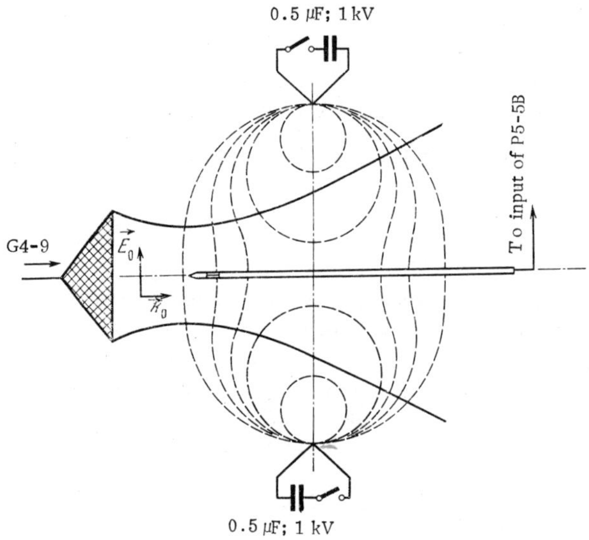

Fig. 1. The location of the probe in the plasma layer. The Langmuir field signal is detected by a P5-5B measurement receiver with a passband of 4 MHz. The dashed curves denote contours of equal plasma density at various times.

1. Experimental Arrangement

The fields of the plasma oscillations and waves were measured on a device which simulates the conditions of free space. This apparatus has been described in sufficient detail elsewhere [17] and here we give only a schematic representation of the main features of the experiment (Fig. 1). In this experiment we have attempted to realize oblique incidence on the plasma boundary with an angle of incidence that is close to optimal. Figure 1 shows that a weakly diverging flux of 10-cm radiation falls on a plasma layer whose critical boundary layer is deformed in time. Thus, if the microwave generator operates continuously or is turned on before the plasma injectors, then the angle of incidence of the beam on the plasma boundary will vary in time and pass through the optimum value.

The plasma layer is produced by four Bostick gun-type spark injectors [20, 21].

A 0.5-μF condenser charged to a voltage of 0.5-5.5 kV is discharged across a gap consisting of a Plexiglas surface in a discharge of about 1 cm in length. The discharge is triggered with the aid of an auxiliary spark along one of the electrodes. The energy of the trigger pulse is less than 0.2 J. In this parameter range it was possible to trigger the guns almost simultaneously with the aid of a trigger circuit containing a TGI-1-325/15 thyratron.

The source design included provision for observing the condenser discharge current. Measurements showed that the discharge is oscillatory and ends completely within a time of 0.2-0.3 μsec.

Injectors of this type have been studied repeatedly [21, 22]. It has been shown experimentally that near the gun a plasma is formed with an electron density on the order of 10^{14}-10^{15} cm^{-3} and expands with a velocity of 5×10^6 cm/sec. At distances of several tens of centimeters from the source the plasma has an electron temperature of T_e = 5-10 eV. The neutral gas released from the dielectric moves at a velocity of 10^5 cm/sec so that over several tens of microseconds the vacuum conditions in the device are practically unchanged.

Proceeding from the experimental data of Gol'ts [21] which show that the electron density distribution in the longitudinal direction obeys a cubic law $n(x) = N/x^3$ at a sufficient distance from the gun, and assuming that the transverse density distribution can be characterized by some inhomogeneity distance a such that

$$n(x, z) = n_m(x) e^{-z^2/a^2},$$

we can compute the density distribution inside the apparatus. The results of such a calculation are shown in Fig. 2 where the constants were determined from the experimental result that the plasma density falls by a factor of 10 when the measurement probe is moved in the transverse direction to the location at which the

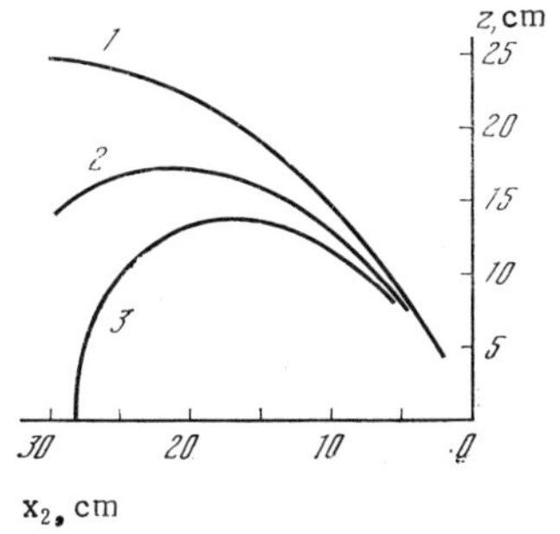

Fig. 2

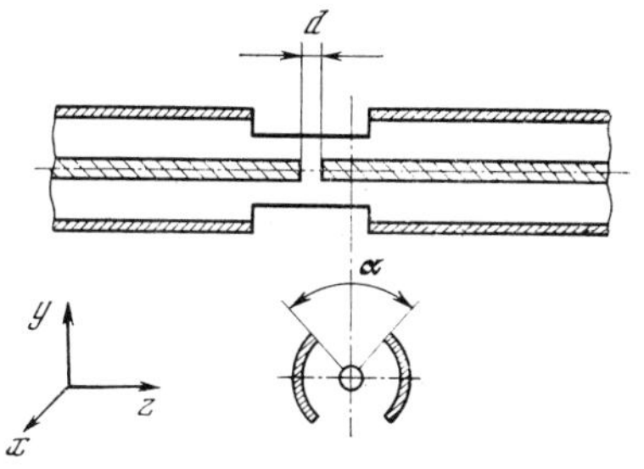

Fig. 3

Fig. 2. The calculated location of the equal density contours in the xz plane for a single source: (1) n = 2.5 × 10^{11}; (2) 7.5 × 10^{11}; (3) 1.5 × 10^{12}.

Fig. 3. The screened probe for detecting the high-frequency Langmuir field.

direction to the plasma source forms an angle of 45° with the initial direction. It is also assumed that the source is able to produce a density $n_m = 10^{14}$ cm^{-3} at the outlet of the pipe. This figure characterizes the width of the computed topographic pattern and demonstrates that it is possible to create a layer with a relatively smooth critical density boundary. The calculated size of the plasma inhomogeneity in the center of the chamber is a = 19 cm, while on the boundary of the localization region of the microwave field (x = 20 cm) a = 13 cm. It should be noted that these numbers characterize the operation of one injector alone rather than the experimental apparatus when it is bounded by the other injector tubes.

In the experiment the plasma layer is produced by four injectors each of which ejects a plasma stream toward the center of the vacuum vessel. The density was measured absolutely by recording the ion current to a single Langmuir probe which could be moved in the vessel.

The density distribution along the z axis in the plasma layer as constructed from the probe readings has a nearly bell-shaped form. A slight asymmetry on the right-hand side is unimportant for the experiment since the main processes occur for $z < 24$ cm.

For exciting and studying longitudinal waves in plasmas it is customary to use probes in the form of a plane collector and grids which excite a coaxial line [23]. The coupling of such a probe with a longitudinal wave is determined by the area of the collector and may be rather high. The use of such probes in high external microwave fields, however, involves some difficulty, primarily because the screening requirements for the probe are greater. When measuring strong fields the coupling of the probe may be reduced; that is, the area of the collector may be reduced. Then the conditions for screening are eased. We avoided screening the probe with a grid since then the external field in the working region of the probe is distorted and the plasma density is somewhat reduced.

In our work the plasma oscillations were studied with a probe that was screened by a length of so-called open coaxial line (Fig. 3). The probe is coupled with the electron oscillations through a gap in the center conductor of the line. The design parameters of the probe noted in Fig. 3 had the values α = 90° and d = 1 mm in the experiment.

The operating principle of the probe is based on the known property of a coaxial line that the maximum energy density in the line is concentrated near the center conductor and decreases with distance from it. This effect is especially noticeable in high-impedance lines. Thus, nonuniformities in the outer conductor have little effect on the structure of the rf wave inside it. In our experiment, this nonuniformity is a longitudinal slit whose width corresponds to an angle α = 90°. It can nevertheless be stated that the rf current induced in the center conductor of the line is strongly coupled with the TEM oscillations in the line.

Thus, if there are plasma oscillations with their wave vector directed along the coaxial axis (**z**) in the plasma near the gap in the center conductor, then these oscillations create an rf current in the center conductor of the line. As a result, a TEM mode is created in the line with an amplitude that is proportional to the field strength in the plasma oscillations. The center conductor of the line can be excited (through the gap) either by the electron current or by the capacitive coupling with the oscillating plasma electrons. It can be shown that when $\omega_0 \approx \omega_{Le}$ the capacitive current is several orders of magnitude greater than the particle current.

It should be noted that the range of wave numbers of the plasma waves that can be detected by the probe extends from the maximum k existing in the plasma ($\sim r_{De}^{-1}$) to extremely small k whose value is determined by the geometry of the screening system.

The shielding of the probe was tested at low power by means of a signal generator. The probe was placed in a rectangular waveguide and the amplitude of the signal which entered the coaxial of the probe because of the distortions in the field and because of imperfections in the open line was measured. When the external field $\mathbf{E}_0$ is parallel to the **x** or **y** axes (see Fig. 3), the screening coefficient is equal to 10^{-3}; when $\mathbf{E}_0$ is parallel to **z**, the screening coefficient is 2×10^{-3}. These values are the ratio of the amplitudes of the signal which penetrates to the signal from the probe without the shielding system.

2. Conditions for Transformation of a Transverse Wave into a Longitudinal Plasma Wave

A number of conditions must be fulfilled for linear excitation of Langmuir waves to occur in an inhomogeneous plasma layer. First, the electric field of the incident wave must contain a component parallel to the density of the plasma (E_z). This requirement is equivalent to having the angle of incidence ϑ of the transverse wave be nonzero. Theory shows that there is some optimum angle for linear transformation $\vartheta_{opt} = (\omega L/c)^{-1/3}$ at which the transformation coefficient $R_{tr}^2 \simeq 0.5$. This coefficient is defined as the ratio of the energy flux removed by the plasma wave to the incident flux. It is obvious from general considerations that 0.5 is the maximum value of the coefficient corresponding to a matched power distribution between reflection and transformation.

We note also that such values of the transformation coefficient are obtained when the approximation of geometric optics is valid, that is, when the size of the inhomogeneity in the layer $L \gg \lambda_0$.

The optimum value of the angle ϑ for our experimental conditions ($L = 10$–20 cm) is 30–24°. Evidently, if the incident wave is turned on at a time when the critical density boundary has not yet reached the field localization region, or, in general, a continuous generator is used, then the transformation coefficient will vary in time since the plasma boundary moves. If $E_0^2/8\pi \ll n\varkappa T_e$ then the experimental geometry is completely determined by the plasma motion and the angle of incidence of the wave must pass through the optimum value during formation of the layer (see the calculated locations of the contours of equal density in Fig. 2).

For oblique incidence of a transverse wave with E_0 in the plane of incidence, energy reflection occurs at the point $\varepsilon = \sin^2 \vartheta$ and the field penetrates to the point $\varepsilon = 0$ (near which the transformation occurs) because of the subbarrier effect and increases there. Transformation can be significant only when, during this increase, the magnitude of the Langmuir field $E_l \gg E_0$.

In theoretical papers there are several different ways of estimating this growth in the field. If the amplification of the field is limited by escape of energy into plasma waves [1, 24], then amplification of the field is roughly $E_l/E_0 \simeq (L/r_{De})^{2/3}$, which yields $E_l/E_0 \simeq 200$ for our experimental conditions.

Other estimates also yield substantial field increases in the neighborhood of the plasma resonance. Thus, when the field is limited by nonlinearity in the electron motion [24], we obtain

$$\frac{E_l}{E_0} = \left(\frac{2\pi e E_0}{m\omega^2 \lambda}\right)^{-2/3} \simeq 50.$$

Therefore, under our conditions the field must increase rapidly in the resonance region.

It is necessary that the effect of electron collisions be sufficiently small for field amplification to take place. The electron–ion collision frequency is

$$\nu_{ei} = \frac{4}{3} \frac{\sqrt{2\pi}\, e^4 n \Lambda}{m^2 v_{Te}^3} = 2.5 \cdot 10^5 \ \text{sec}^{-1} \ (\Lambda = 10),$$

so that the imaginary part of ε in the resonance region is $\varepsilon'' = \nu_{ei}/\omega_{Le} \simeq 10^{-5}$, which corresponds to extremely low absorption. The frequency of electron collisions with the wall of the vacuum chamber (more precisely, with the plasma boundary alongside it) is somewhat higher than ν_{ei} and is equal to

$$\nu_W = v_{Te}/L_W = 10^8/60 = 1.7 \cdot 10^6 \ \text{sec}^{-1};$$

however, in evaluating the field amplification, ν_W may be neglected since field amplification occurs in a resonance region of size $\Delta L \sim 1$ mm while the distance covered by a thermal electron during the period of the external field is

$$v_{Te}/f_0 = 10^8/(3\cdot 10^9) = 0.03 \text{ cm}.$$

Consequently, the motion of the thermal electrons is adiabatic within the field localization region and they do not remove energy from the growing E_l field. Only the Maxwellian tail of the electrons with speeds

$$v > \Delta L \omega_0 = 2\cdot 10^9 \text{ cm/sec} = 20\, v_{Te}$$

can remove energy from the localized E_l field, and damping caused by these electrons is certainly small compared to collisional (via ν_{ei}) absorption.

3. Deformation of the Field Distribution near the Plasma Resonance Region in the Presence of Small but Finite-Amplitude Fields

The spatial distribution of the Langmuir field in the neighborhood of the plasma region was studied experimentally. The incident electromagnetic wave was produced by a low-power signal generator (G4-9) operating in a continuous mode. The operator power was controlled within 25 dB and the maximum power (30 mW) corresponded to a dimensionless field of $E_0/E_p = 10^{-3}$. Unlike experiments with a high-power generator, in this experiment the interaction between the microwave field and the plasma begins practically from the moment the $n = n_c$ boundary reaches the region of field localization as the plasma moves from the sources.

Since in this case the angle of incidence of the wave on the plasma is close to the optimum for linear transformation, the Langmuir field at the point $\varepsilon \simeq 0$ begins to be amplified and reaches maximum values of $E_l = 2\cdot 10^{-1}\, E_p$ (under the experimental conditions for $P_0 = 30$ mW). Here the Langmuir waves propagate into the rarefied plasma region.

As the probe was located at a fixed point on the axis of the chamber between the transmitter horn and the plane of the injectors (although it could be moved along the chamber axis), in order to extend the Langmuir field signal as much as possible in time, the following procedure was followed. The voltage on the injectors was chosen so that the critical density boundary moved as close as possible to the probe without crossing it and then moved backward because of thermal expansion (in general with a lower velocity as the plasma temperature is slightly lower). Therefore, the E_l field or the l-wave are detected in an underdense plasma region. The exact location of the point $\varepsilon = 0$ relative to the plasma, however, is unknown since adjustment to this regime was done from the shape of the l-signal. If the dense plasma crossed the probe, then two broad maxima were seen in the signal (Fig. 4). The regime with the largest signal but only one maximum was chosen for the measurements.

We note that in this series of measurements the degree of homogeneity of the layer L was enhanced slightly by balancing the injector currents. Measurements showed that $L = n/\nabla n = 20$ cm.

Therefore, for negligibly small values of the incident wave field the E_l field signal has a single broad maximum corresponding to the closest approach of the critical density boundary to the probe (Fig. 5a). When $P_0 \geq 1$ mW ($E_0/E_p \geq 2\cdot 10^{-4}$) the shape of the signal is strongly distorted, which is also an indication that the spatial distribution of the E_l field is changed. At the same time, the averaged (over many shots) signal amplitude over the entire range of powers studied in this series of measurements is proportional to the amplitude of the incident wave to within ±2 dB. At the maximum power ($E_0/E_p = 10^{-3}$) the peaks in the E_l signal become narrower and less distinct.

We now point out that (see the oscilloscope traces of Fig. 5) dissipation of the Langmuir wave may possibly increase when the field distribution is deformed. In fact, since the average width of the signal becomes smaller and the E_l field appears in the rarefield plasma region (far from the point $\varepsilon \simeq 0$) only because of propagation of an l-wave from the excitation point, it is apparent that some absorption mechanism is involved in this propagation.

4. Nonlinear Damping of Propagating Langmuir Waves

Measurements were also made with a high-power microwave generator to study the propagation of Langmuir waves. Since the field pressure may then approach $n\varkappa T_e$ (and, as shown in [17], the dense plasma does not usually reach the center of the chamber if the generator is turned on beforehand), the technique of the

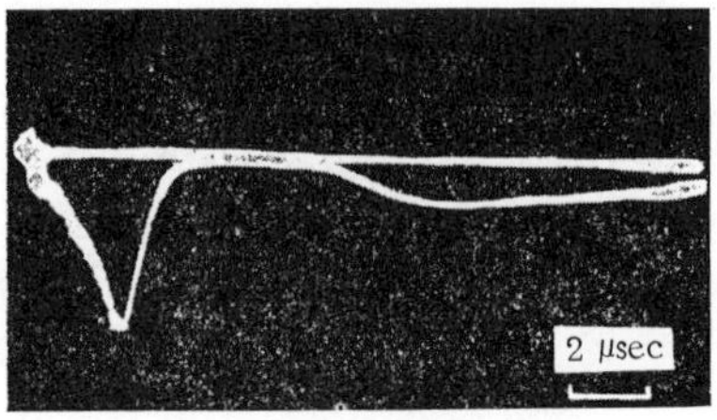

Fig. 4

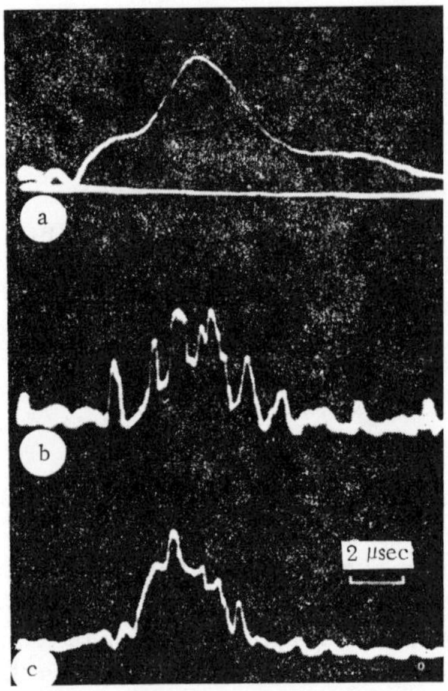

Fig. 5

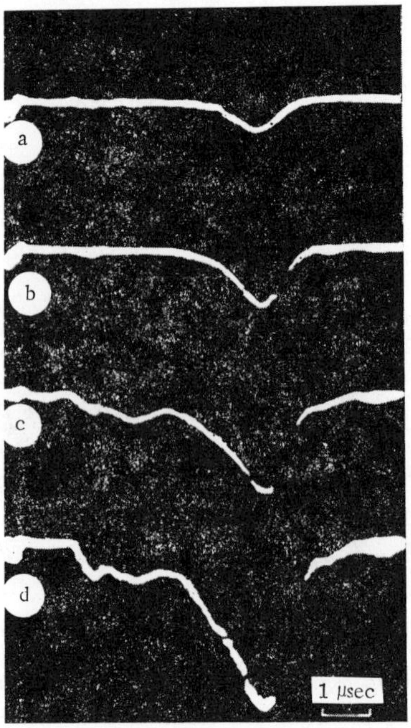

Fig. 6

Fig. 4. A comparison of the Langmuir field signals at the leading and trailing edges of the plasma. A delay of 4 μsec has been introduced from the time the guns are turned on. Detection is linear.

Fig. 5. The Langmuir field signals. The spark sources were fired 10 μsec before the maximum in the Langmuir signal. Quadratic detection was used. Amplification: (a) 1 dB; (b) less than 15 dB; (c) less than 25 dB. Power: (a) 0.1 mW; (b) 3 mW; (c) 30 mW.

Fig. 6. The change in the Langmuir field signal as the power of the incident wave is varied at high fields: (a) $E_0 = 60$ V/cm; (b) 120 V/cm; (c) 240 V/cm; (d) 360 V/cm.

experiment was changed somewhat. The generator was turned on at a time when the layer had formed but the critical density boundary had not yet been smoothed out; consequently, the conditions for linear transformation were satisfied.

The maximum plasma density (in the interior of the layer) was chosen so that the probe was reliably blocked. This was detected from the cutoff of the E_l field signal since the Langmuir oscillations as well as the transverse waves are screened out by an overdense plasma. For a fixed location of the probe, the E_l signal has two maxima (see Fig. 4), of which the first corresponds to passage of the region with $n \simeq n_c$ past the probe as the boundary moves from the injectors to the probe, while the second corresponds to motion of the boundary in the opposite direction. In this case the change in the velocity of the plasma boundary already shows up strongly and is reflected in the oscilloscope traces as a spreading out in time of the second maximum. The amplitude of the signal during the reverse motion of the boundary in the plasma layer is down by roughly a factor of 4, which may be explained by the change in the angle of incidence and its approach to zero as the plasma decays.

The dependence of the E_l signals on the incident wave power can easily be traced in a set of oscilloscope traces (Fig. 6) for the leading edge of the plasma. The range of variation of E_0/E_p in this case is 0.05-0.3. We note that in this range of fields the deformation of the resonance region (if it even exists) should not be detected because of the low temporal resolution. The average pattern of the Langmuir field is measured and its form at low fields corresponds to that calculated for collisionless Landau damping of the waves. A calculation according to the formula

$$\tilde{\gamma}_0 = \sqrt{\frac{\pi}{8}} \frac{\omega_{Le}^4}{k^3 v_{Te}^3} \exp\left(-\frac{\omega_{Le}^2}{2k^2 v_{Te}^2} - \frac{3}{2}\right)$$

demonstrates that damping of the waves by a factor of e^2 should occur when the plasma density falls to $0.8n_c$. In fact, the detected E_l signal for $E_0/E_p = 0.05$ ($E_0 = 65$ V/cm) lies in a plasma density range from $0.7n_c$ to

$n \simeq n_c$ as can be seen from Fig. 6a. In this and later calculations it is assumed that the density distribution in the layer is close to linear. In addition, it is understood that the value of v_{Te} in the plasma does not change during the time of the signal and, therefore, the time scale in Fig. 6 corresponds linearly to the change in the plasma density.

Increasing the power of the incident wave and, thus, the amplitude of the E_l fields causes the signal to broaden considerably in the underdense plasma region. Experiments show that for $E_0/E_p = 0.3$ practically the entire region from n_c to $^1/_4 n_c$ is filled by the field of the electron plasma waves.

5. Discussion of Results

Normal and Oblique Incidence of the Waves

The angle of incidence plays an important role in the interaction of a directed flux of high-power electromagnetic waves with an inhomogeneous plasma layer. The transformation of a transverse wave into a longitudinal plasma wave, which does not occur for strictly normal incidence, is described in general for moderate angles of incidence $\vartheta \lesssim (\omega L/c)^{-1/3}$ by the formula for the transformation coefficient defined as the ratio of the energy fluxes in the waves [25]:

$$R_{tr}^2 = \frac{16\pi n \varkappa T_e}{E_0^2} |\varepsilon_L| \frac{v_{Te}}{c} \left(\frac{M}{m}\right)^{1/2} \sin^2 \vartheta.$$

In this expression ε_L is the dielectric constant at the point to which the wave penetrates in the steady state. The formula is valid for sufficiently large fields such that

$$\frac{E_0^2}{8\pi n \varkappa T_e} \gtrsim 4\pi \frac{v_{Te}}{c} \frac{r_{De}}{L},$$

which correspond to $E_0/E_p \gtrsim 4 \cdot 10^{-3}$ for the experimental conditions ($v_{Te} = 10^8$ cm/sec; $r_{De} = 5 \cdot 10^{-3}$ cm; $L = 15$ cm). It is evident that a large fraction of the energy of the incident wave can be converted to plasma waves, a fact which is important for explaining the dissipation of the incident wave and its anomalous propagation.

Nonlinear Effects in the Field Distribution of the Plasma Waves

Our measurements of the longitudinal field in the neighborhood of the plasma resonance can be compared with theoretical ideas on the deformation of the resonance region. We have found a strong distortion of the field distribution at incident wave fields $E_0/E_p > 2 \cdot 10^{-4}$. It is important to note that such small energy fluxes (the pressure of the transverse wave field is 10^{-6}–10^{-8} of $n\varkappa T_e$) are unable to change the average motion of the plasma in a large volume (on the order of λ_0^3).

A local change in the field (and plasma density), however, does not require large expenditures of energy. For example, in work on the distributions of the field in inhomogeneous collisional plasmas with very large sized inhomogeneities [26, 27] field amplification of up to 1000 times and the formation of "dips" in the plasma density have been observed. These effects were observed when $E_0^2/(8\pi n \varkappa T_e) = 10^{-6}$–$5 \cdot 10^{-5}$. It may be assumed that under our conditions, isolated regions with a trapped field are also formed; however, many such regions are generated. Evidently this effect must be explained by noting that in the initial stage the deformation of the plasma resonance region must play an important role [28]. This deformation appears in a developed state over a time $\Delta L/v_s$ ($v_s = \sqrt{\varkappa T_e/M}$ and ΔL is the size of the resonance region). In these experiments $v_s = 2.3 \times 10^6$ cm/sec, which exceeds the directed velocity of the plasma (10^6 cm/sec); thus, a nearly steady state is established during the time the plasma boundary moves from the sources to the probe.*

It follows from the theory that the condition $\eta = (D^2/E_p^2)(L^2/r_{De}^2) > 1$ (D is the electric induction) must be satisfied for strong deformation to occur. If we assume for an estimate that $D = E_0$, which is true for the optimum angle of incidence and for $L \sim \lambda_0$, then we obtain the rather close value $\eta = (2 \times 10^{-4})^2 (20/0.005)^2 = 0.65$. We can also use the fact that in our experiment some amplification of the field E_0 is possible in the center of the layer because of reflections caused by the transverse inhomogeneity,

*G. M. Fraiman brought attention to the importance of this question. In fact, our experiment does not provide an answer to the question of whether this pattern of deformation of the field is conserved as time increases. The necessary condition for establishing a steady state is, however, satisfied.

It is of interest to estimate the spatial scale of the inhomogeneity in the E_l field. In the limit of infinitesimal amplitudes $\Delta z_0 = (r_{De}^2 L)^{1/3} \simeq 1$ mm; however, Δz_0 was not measured in our experiments since the maximum in the signal is obtained when the boundary stops near the probe (evidently, the critical density point does not reach the probe, otherwise the signal would have two maxima).

Experimentally, we can only estimate Δz during the deformation. Assuming that the velocity of the boundary is about 5×10^5 cm/sec (from the measurements of [17]) and measuring the width of the field peaks far from the maximum, we obtain $\Delta z \leq (4 \times 10^{-7}) \times (5 \times 10^5) = 2$ mm, that is, Δz has the same order of magnitude as Δz_0, about $20 r_{De}$. As the incident wave power is increased the characteristic scale length of the field oscillations must decrease (as $\eta^{-1/6}$). A narrowing of the peaks in the maximum field is also observed experimentally and they cease to be observed since the duration of the signals becomes less than the temporal resolution of the apparatus.

The amplitude of the deformation in the field, however, is considerably larger than that predicted by theory (even for $\eta = 20$). For a stationary state it is evident that there must be equality between the pressures of the deformed part of the field and the plasma density perturbation, i.e.,

$$\Delta n \varkappa T_e = \Delta E_l^2 / 8\pi.$$

Thus, the theoretically predicted deformation $\Delta\varepsilon/\varepsilon$, on the order of 1%, should correspond to $\Delta E_l / E_l = 0.1$. Field deformation of up to 0.5 is observed experimentally with a quadratic detector, that is, a field $\Delta E_l / E_l \simeq 0.7$.

We point out that in the rarefied plasma region the intensity of the electron plasma wave is reduced in the presence of deformations. This is apparently related to the dissipation of energy. It is of course possible that this effect is due to reflection (of a plasma wave); however, then this should lead to a change in the linear transformation coefficient R_{tr} and to an increase in the reflection coefficient of the transverse wave, which was not observed in our experiments.

The Reduction in Collisionless Damping

Our results are in qualitative agreement with Vedenov's formula for the reduction in damping, $\tilde{\gamma} = \tilde{\gamma}_0 / (1 + AE^{3/2})$ [7]. In general, the formula for Landau damping γ_0 ceases to be valid when (see [1]

$$\frac{eE_l}{k} \simeq \left(\frac{12\pi e^4 n \varkappa T_e}{k m^{3/2} v_{ph}^3} \right)^{2/3}, \quad \text{where } v_{ph} = \frac{\omega_l}{k}.$$

Under our experimental conditions, for $k \sim 1/r_{De}$ the damping of waves in the rarefied plasma is already decreasing when $E_l = 1$ V/cm.

Here the qualitative picture is rather simple. During propagation of a wave into a rarefied plasma the wave number k increases in accordance with the dispersion relation $\omega_0^2 = \omega_{Le}^2 + 3k^2 v_{Te}^2$. This corresponds to a reduction in the phase velocity of the wave so that an ever larger number of electrons falls into synchrony with the wave. Consequently, a stronger wave will be absorbed by a larger number of electrons; that is, it will reach a region of more rarefied plasma. The limiting value for $k = r_{De}^{-1}$, when substituted into the dispersion relation, gives the minimum density $n \sim {}^1/_4 n_c$ at which the concept of electron plasma wave still has meaning. We note that reduced damping with increasing fields has also been observed within limits in [29], where enhanced damping was observed when the field was increased further.

The question of whether in general an l-wave will be damped if it is sufficiently strong is important. The criterion for strong damping is the condition that the potential energy of an electron in the wave, $eE_l \lambda_l$, be less than $\varkappa T_e$. For the maximum field in this experiment ($E_0 = 360$ V/cm) with $k \simeq r_{De}^{-1}$, we find that damping continues to be large since $\varkappa T_e / e r_{De} \simeq 1150$ V/cm $\gg E_0$. Thus, the wave cannot propagate in the density region where $n/n_c < 1/4$. This is in agreement with the measured interval of plasma densities over which an E_l field was detected.

Conclusions

The basic results obtained here can be summarized as follows.

An experimental study of the field distribution of the electron plasma waves shows that, for oblique incidence of a wave in which the field exceeds the level $E_0/E_p > 2 \times 10^{-4}$, strong deformation of the E_l field distribution occurs in the region where the plasma density n is close to (but less than) the critical density.

Narrow (on the order of 2 mm) peaks in the field distribution are detected against a background of the broader E_l signal corresponding to negligibly small fields.

When the power is greatly increased (to $E_0/E_p > 6 \times 10^{-2}$) the signal from the E_l field expands strongly into the rarefied plasma region and is detected at points where it is calculated that $n \simeq {}^1/_4 n_c$.

LITERATURE CITED

1. V. L. Ginzburg, The Propagation of Electromagnetic Waves in Plasmas [in Russian], Nauka, Moscow (1967).
2. V. E. Golant and A. D. Piliya, Usp. Fiz. Nauk, 104:413 (1971).
3. A. D. Piliya, Zh. Tekh. Fiz., 36:818 (1966).
4. V. P. Silin, The Parametric Interaction of High-Power Radiation with Plasmas [in Russian], Nauka, Moscow (1973).
5. A. A. Galeev and R. Z. Sagdeev, Reviews of Plasma Physics, Vol. 7, Consultants Bureau, New York (1978).
6. N. S. Erokhin and S. S. Moiseev, ibid., p. 146.
7. A. A. Vedenov, Reviews of Plasma Physics, Vol. 3, Consultants Bureau, New York (1967).
8. T. F. Volkov, Plasma Physics and the Problem of Controlled Thermonuclear Reactions [in Russian], Vol. 3, Izd. AN SSSR, Moscow (1958), p. 336.
9. A. A. Vedenov and L. I. Rudakov, Dokl. Akad. Nauk SSSR, 159:767 (1964).
10. V. E. Zakharov, Zh. Éksp. Teor. Fiz., 62:1745 (1972).
11. A. Y. Wong, R. L. Stenzel, H. C. Kim, and F. F. Chen, V International Conf. on Plasma Physics and Controlled Thermonuclear Fusion Research, Tokyo (1974), paper IAEA-CN-33/H4-1.
12. H. Ikezi, K. Nishikawa, H. Hojo, and K. Mima, ibid., IAEA-CN-33/H4-3.
13. G. M. Batanov, K. A. Sarksyan, and V. A. Silin, Plasma Diagnostics [in Russian], Vol. 3, Atomizdat, Moscow (1973), p. 513; Fiz. Inst. Akad. Nauk SSSR Preprint No. 92 (1970).
14. V. A. Silin, XII International Conf. on Phenomena in Ionized Gases, Eindhoven (1975), Vol. 1, p. 314; Fiz. Inst. Akad. Nauk SSSR Preprint No. 33 (1975).
15. I. R. Gekker and O. V. Sizukhin, Pis'ma Zh. Éksp. Teor. Fiz., 9:408 (1969).
16. G. M. Batanov, K. A. Sarksjan, and V. A. Silin, IX International Conf. on Phenomena in Ionized Gases, Bucharest (1969), Vol. 1, p. 541.
17. Microwave–Plasma Interactions (Proceedings of the P. N. Lebedev Physics Institute, Vol. 73), Consultants Bureau, New York (1975).
18. Yu. Ya. Bordskii, B. G. Eremin, A. G. Litvak, and Yu. A. Sakhonchik, Pis'ma Zh. Éksp. Teor. Fiz., 13:163 (1971).
19. G. M. Batanov and V. A. Silin, Pis'ma Zh. Éksp. Teor. Fiz., 14:445 (1971).
20. W. H. Bostick, Phys. Rev., 106:404 (1957).
21. É. Ya. Gol'ts, Candidate's Dissertation, Physics Institute, Academy of Sciences of the USSR (1968).
22. É. Ya. Gol'ts, V. B. Turundaevskii, and A. Z. Khodzhaev, Fiz. Inst. Akad. Nauk SSSR, No. 129 (1968).
23. G. Van Hoven, Phys. Rev. Lett., 17:169 (1966).
24. S. V. Bulanov and L. M. Kovrizhnykh, Fiz. Plazmy, 1:1016 (1975).
25. R. Z. Sagdeev and V. D. Shapiro, Zh. Éksp. Teor. Fiz., 66:1651 (1974).
26. R. L. Stenzel, A. Y. Wong, and H. C. Kim, Phys. Rev. Lett., 32:654 (1974).
27. H. C. Kim, R. L. Stenzel, and A. Y. Wong, Phys. Rev. Lett., 33:886 (1974).
28. V. B. Gil'denburg and G. M. Fraiman, Zh. Éksp. Teor. Fiz., 69:1600 (1975).
29. R. N. Franklin, S. M. Hamberger, and G. J. Smith, V Europ. Conf. on Controlled Fusion and Plasma Physics, Grenoble (1972), Vol. 1, p. 132; Vol. 2, p. 243.

A STUDY OF SECONDARY-EMISSION MICROWAVE DISCHARGES WITH LARGE ELECTRON TRANSIT ANGLES

L. V. Grishin, A. A. Dorofeyuk,
I. A. Kossyi, G. S. Luk'yanchikov,
and M. M. Savchenko

The conditions for production of a secondary-emission microwave discharge in systems with a large distance between the walls (cavity resonators, waveguides, etc.) and in free space are analyzed. The traditional theory of a resonance discharge is shown to be invalid for these cases and a new method is proposed for determining the thresholds for the discharge. The discussion takes into account the contribution to the discharge current from electrons emitted at all phases of the field. The theoretical calculations are compared with experimental results.

INTRODUCTION

The interaction of microwaves with metallic and dielectric surfaces in a vacuum is described in the framework of classical electrodynamics up to a certain, fairly high-power level. Above some value of the microwave field the character of the wave interaction changes sharply – an electron avalanche develops at the surface of a solid as a result of secondary emission from the surface in the microwave field.

The range of problems in physical research and technology in which one encounters avalanche-type secondary-emission discharges (SED) is extremely wide. Thus, for example, it is the formation of electron avalanches at the transparent output windows of microwave oscillator tubes which limits the attainable power. Discharges in waveguides and circuit elements prevent the transport of high-power electromagnetic waves. An SED can make difficult the operation of rf communication and radar equipment located on objects in space and operating under high-vacuum conditions.

The SED is also well known in accelerator technology, where it is a major factor in the disruption of normal operation of linear and circular charged-particle accelerators.

Secondary-emission breakdowns have appeared in a promising research area for controlled thermonuclear fusion, the heating of plasmas by microwave radiation. The promise of this method is limited by the fact that attempts to increase the microwave power delivered to a plasma device inevitably lead to electron breakdowns on the input windows, in the waveguide circuits, and in the vacuum chamber itself.

Finally, we must note an extremely important circumstance, namely that secondary-emission discharges may make it much more difficult to interpret many experiments, particularly those dealing with the interaction between high-power microwaves and plasmas [1-3].

Efforts to describe the mechanism for formation of the electron cloud have led to the development of the theory of the so-called secondary-emission resonance discharge (SERD).

The traditional model assumes that the conditions under which a certain group of electrons (the resonance electrons) can undergo cyclical motion are satisfied [4-6]. The term cyclic is taken to mean that the electrons appear in free space for a time $t_f = (\pi/\omega)(2k-1)$ (in the case of a discharge between two surfaces) or $t_f = (2\pi/\omega)k$ (for a discharge on an isolated surface). Here t_f is the time interval from the moment an electron escapes the surface until it returns to it or until it reaches an opposing surface; k is the order of the resonance; and ω is the angular frequency of the microwaves. If the trajectories of the resonance electrons are stable and the energy at which the electrons collide with the surface exceeds the first critical potential (the primary electron energy at which the secondary emission coefficient $\sigma = 1$), then continuous multiplication of the electrons may occur with formation of an electron avalanche.

The theory of the secondary-emission resonance discharge has been verified by special experiments. However, there is a wide range of phenomena which occur in high-power microwave fields where strict resonance conditions are clearly not fulfilled. This paper is devoted to developing the theory of this type of "nonresonance" discharge, as well as to setting up experiments for examining the mechanism by which microwave electron avalanches occur in vacuum at the surface of isolated solids under nonresonance conditions.

CHAPTER I

A THEORETICAL STUDY OF THE CONDITIONS FOR A "NONRESONANT" DISCHARGE

1. Role of the "Thermal" Velocity Spread of the Electrons in a Secondary-Emission Microwave Discharge

Theoretically it is convenient to examine the secondary-emission discharge under geometrical conditions in which a variable electric field perpendicular to the surface is located in the gap between two parallel infinite plates and acts on an electron with a force $F \sin \omega t$. The equations of motion of an electron are found by integrating the expression

$$m\ddot{x} = F \sin \omega t, \tag{1}$$

where m is the electron mass, x is the coordinate normal to the surface, and t is the time, to yield

$$\dot{x} = -v_m \cos \omega t + v_m \cos \omega t_1 + v_0, \tag{2}$$

$$x = \frac{v_m}{\omega}(-\sin \omega t + \sin \omega t_1) + v_m \cos \omega t_1 \cdot (t - t_1) + v_0 (t - t_1). \tag{3}$$

Here $v_m = F/m\omega$. Equations (2) and (3) have been obtained under the assumption that $x = 0$ and $\dot{x} = v_0$ at the time t_1 when the electron is emitted from the surface.

Attempts to describe the process whereby electron avalanches appear in such systems have led to the development of the theory of the SERD. According to this theory, all the electrons emitted at time t_1 have the same initial velocity v_0 and are concentrated in a thin sheet parallel to the emitting surface.

A collision with the other surface occurs after a flight time of

$$t_k = T(k - {}^1/_2), \tag{4}$$

and in a one-sided discharge, after a time

$$t_k = Tk, \tag{5}$$

where k is the order of the resonance.

The idea of the discharge having such a resonance character has by now become traditional. However, in the following discussion we shall show that the concept of a "thin sheet" and, therefore, the SERD theory, is valid only to a very limited degree.

The basis of the following considerations is the experimental fact that the bulk of the secondary electrons have a spread in velocity v_0 which is close to Maxwellian at an electron temperature of $T_e \sim$ 2-4 eV. This distribution depends only weakly on the energy of the primary electrons [7].

Unlike in the idealized model of a "thin sheet," the real electrons emitted in phase will be smeared out more and more during the time of flight because of the Maxwellian spread in their thermal velocities v_0 and this smearing will increase with the time of flight.

If we begin with the nature of the density distribution of the in-phase particles in the space in front of the surface on which the discharge occurs, then three types of discharge may be identified: resonant, polyphase, and uniform polyphase.

1. Resonant Discharge. The bulk of the in-phase particles, as is required by the SERD theory, are concentrated in the plane of the "thin sheets." The thickness l of a "thin sheet" is much less than the distance between the sheets, i.e.,

$$l \ll L. \tag{6}$$

We shall define the thickness of a "thin sheet" as the distance between the in-phase electrons which receive a large "thermal" addition to their velocity $v_0 = v_2$ and the electrons which receive a small addition $v_0 = v_1$ [v_1 and v_2 are defined by the half-width of the "thermal" speed (v_0) distribution curve of the secondary electrons]. By substituting in Eq. (6) the values $l = (v_2 - v_1)t_f$ and $L \simeq v_m \cos \omega t_1 \cdot T$, where $t_f = t - t_1$ is the time of flight until a collision, it is possible to find the limits of applicability of the "thin sheet" concept and, therefore, of the SERD theory, imposed by the spread in the "thermal" speeds of the electrons:

$$t_f/T \ll v_m \cos \omega t_1/(v_2 - v_1). \tag{7}$$

For T_e = 2-4 eV, $v_2 - v_1 \simeq 10^6$ m/sec. From the SERD theory it follows that a resonant discharge is possible in terms of stability only when emission phases in the range $0 < \omega t_1 < 32°$ are used [5]. Thus, we can set $\cos \omega t_1 \simeq 1$.

If we replace the sign $\ll$ by $<$ and divide the right-hand side by 10 in the inequality (7), then, using Eqs. (4) and (5) we find the limits of applicability of the SERF theory:

$$k < 10^{-7} v_m + {}^1/_2 \tag{8}$$

for a discharge with two surfaces and

$$k < 10^{-7}\, v_m \tag{9}$$

for a single surface.

From Eqs. (8) and (9) it follows that for electrons with energies of hundreds of electron volts ($v_m \sim 10^7$ m/sec), as is the case in most experiments, the SERD theory yields an accurate picture of the discharge only for the first two resonances, and that is "stretching it." This is in agreement with numerical simulations [8, 9], as well as with a number of experimental papers [10-13], in which the electrons in the resonant discharge were not monoenergetic. Since only one phase of the field contributes to multiplication of the electrons, this discharge may be referred to as a monophase discharge.

2. Polyphase Discharge. If

$$l \sim L, \tag{10}$$

then electrons which fall on the surface in a wide range of phases can contribute to the discharge. Consequently, this type of discharge may be referred to as polyphase. At present there is no theory for it.

3. Uniform Polyphase Discharge. For a time of flight still greater than in the case of a polyphase discharge or for a still smaller v_m, the condition

$$l \gg L \tag{11}$$

will be satisfied.

This condition signifies a uniform distribution of the density of in-phase electrons in front of the secondary-emission surface. We shall refer to this type of discharge as a uniform polyphase discharge. Substituting the values l and L in Eq. (11) and setting $t_f \simeq d/v_m$ (d is the distance between the surfaces), we obtain

$$d \gg 10^{-6}\, v_m^2\, T. \tag{12}$$

When condition (12) is fulfilled, if an SED occurs it will be uniform polyphase.

While the terms polyphase and uniform polyphase are not generally accepted, it seems to us that they reflect most fully the essence of the phenomena being described. We shall use these terms throughout the following discussion.

From the above qualitative analysis, it seems appropriate to use the criteria (6), (10), and (11) [19-21] for determining the regions in which a model is applicable; they are more accurate than the criteria proposed in [14].

2. Analysis of a Uniform Polyphase SED on Materials with a Linear Dependence of σ on the Energy (E ⊥ S)

It is evident that, in determining the threshold for appearance of a polyphase SED in terms of the level of the rf field, it is necessary to include the dependence of the secondary emission coefficient on the electron energy over the entire range of energies which can be acquired by an electron in a given microwave field. This differs from the theory of a resonant discharge, where the secondary emission properties of the material are characterized by a single variable W_1/e, the first critical potential of the substance (e is the electronic charge). This circumstance creates an impression of difficulty in constructing an analytic theory of polyphase discharges. In fact, for a given microwave field at the surface of different materials, they will have different electron distribution functions during a discharge because of their individual energy dependences for σ. Moreover, even in the case of discharges on a single material, the distributions will be different for each new value of the microwave field intensity. It thus seems natural to use the approach proposed in [8, 9], where the distribution is sought by numerical calculation for each particular case, or in [15], where an energy distribution function of the electrons (which corresponds, for example, to a rarefied plasma in the discharge gap) is chosen in advance while the real function is unknown.

However, in a special case of a polyphase discharge – the uniform polyphase SED – the problem is greatly simplified so that it is possible to find the true electron distribution function analytically and then use it to find the criteria for development of a discharge.

Statement of the Problem

In this section the following problem is set: in the initial stage of the discharge, when the effect of space charge can be neglected, find the distribution function of the electrons with respect to the phases of emission and determine the threshold field in a discharge between two infinite planes which lie in an alternating electric field perpendicular to the surface which acts on an electron with force $F \sin \omega t$. The problem is solved under the following assumptions:

1. The time between the moment an electron is emitted (t_1) and the moment a collision occurs with the second wall is so large that the electrons emitted in any phase $\varphi_1 + 2\pi k$ (k is an integer and $\varphi_1 = \omega t$) can be regarded as uniformly spread out in the space in front of the surface which they are to strike. It is easy to show that the condition for a uniform distribution of the in-phase electrons reduces to

$$t_f/T \gg v_m/\Delta v. \tag{13}$$

Here $T = 2\pi/\omega$; Δv is the half width of the distribution of initial speeds of the in-phase electrons and is given by $\Delta v = 10^6$ m/sec for $T_e = 2\text{-}4$ eV; $v_m = F/m\omega$; m is the electron mass; and, t_f is the time of flight for an electron between the walls. This condition will be satisfied, for example, in a standard centimeter-band waveguide at microwave powers of 0.1-1 MW, if we consider a two-sided discharge between the walls or a one-sided discharge on any circuit element where forces are present which return any escaping electrons (e.g., an electrostatic force). It is easy to show that if the time of flight of an electron in the returning field satisfies condition (13), then the one-sided discharge is analogous to the two-sided one.

2. We shall assume that the dependence of σ on the collision energy W_c has the form

$$\sigma = W_c/W_1, \tag{14}$$

where W_1 is the electron energy for which $\sigma = 1$. This dependence is close to the initial part of the secondary emission characteristics for most substances.

3. In determining the dynamics of an electron in front of the discharge surface, we neglect its initial velocity* and assume that the electron's motion is completely determined by the microwave field. This is justified since the bulk of the electrons in a near-threshold field must have an energy $W_c \sim W_1 \gg 2\text{-}4$ eV.†

Equations of Motion

By integrating Eq. (1) (the x axis is taken perpendicular to the emitting surface and parallel to F) and including assumption No. 3, we obtain expressions for the velocity of the electron flux to the wall and for the dimensionless coordinate $\xi = xm\omega^2/F$ of an arbitrary electron:

$$\dot{x} = \frac{F}{m\omega}(-\cos\varphi + \cos\varphi_1), \tag{15}$$

$$\xi = -\sin\varphi + \varphi\cos\varphi_1 + c. \tag{16}$$

Here $\varphi = \omega t$. Since the in-phase electrons are uniformly distributed, c may have an arbitrary value.

*It is, however, just the scatter in the initial velocities which leads to a balancing in space of the density of the in-phase electrons in the x direction.

†In this work we were unable to include the contribution of reflected electrons which usually form 10-20% of the total number of secondary electrons [7]. It may be assumed that their role in the discharge is not very large since, although they have a higher σ because of their large velocity, the fast particles have a larger probability, as will be evident later, of falling in the time of the "inactive" phase of the field and not having any "descendants." This question, however, requires special consideration.

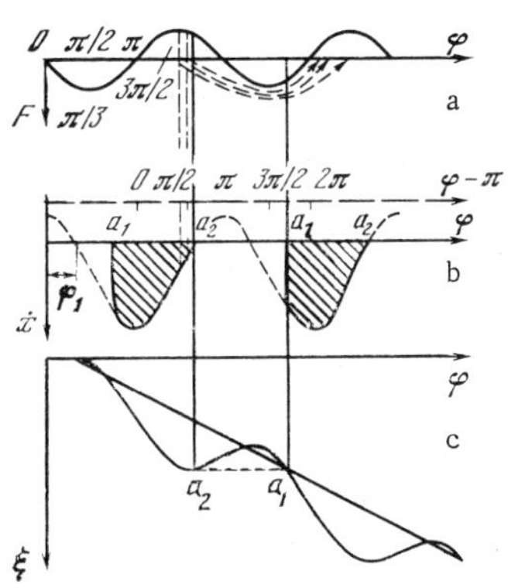

Fig. 1

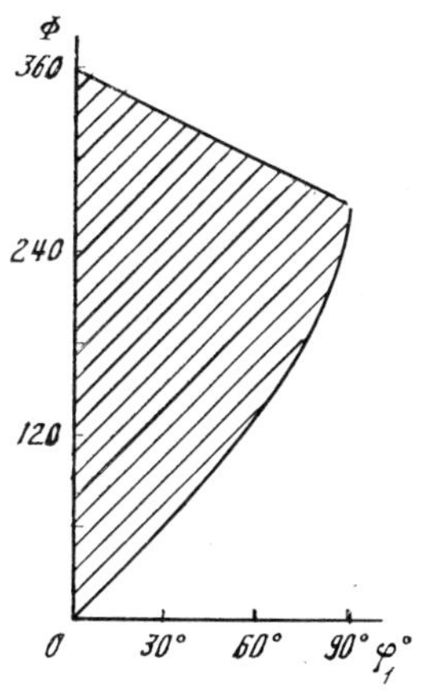

Fig. 2

Fig. 1. The force acting on an electron (a), and the velocity (b) and dimensionless coordinate (c) of an electron emitted with phase $\varphi_1 = 1.05$. The flux density of the in-phase electrons into various phases of the field is shaded.

Fig. 2. The range of phases of the field Φ (shaded) during which an electron can encounter the wall, as a function of the emission phase φ_1. An electron emitted in phase 0 can encounter the wall during an arbitrary phase of the field.

Operating Range of Phases for the Primary and Secondary Currents

It must first be noted that only those electrons emitted by the wall in the range $\varphi_1 = (0-\pi/2) + 2\pi k$ (Fig. 1a) can leave it and participate in the discharge since they have a constant component of velocity $V_t = (F/mw) \times \cos\varphi_1$ directed away from the wall. For compactness the term $2\pi k$ will be left out in the following when determining the phase.

We shall find the range of phases during which an electron may hit an obstacle in its path. The dependence of the coordinate of the electron on the phase of the field according to Eq. (16) is shown in Fig. 1c. As an example we have chosen an electron with a constant velocity component corresponding to an emission phase of $\pi/3$. From the figure it is clear that beginning with a phase of a_2, when the velocity of the electron is 0 (Fig. 1b), and until phase a_1, no electron with this constant velocity can strike an object lying in the x direction. At other phases of the field, the number of particles which come to an opposing surface is proportional to the electron velocity (15) (Fig. 1b). It is easy to see that $a_2 = -\varphi_1$. We can find the point a_1 from Eq. (16), noting that the coordinates of electrons with phase a_1 and phase a_2 coincide, i.e., $\xi(a_1) - \xi(a_2) = 0$. As a result we obtain an equation for finding the point a_1:

$$\sin a_1 - a_1 \cos \varphi_1 + \varphi_1 \cos \varphi_1 - \sin \varphi_1. \tag{17}$$

From a numerical solution of Eq. (17) it is easy to find that electrons emitted in the phase interval 0-1.35 will have $a_1 < \pi$. Evidently, for these particles $a_2 > {}^3/_2\pi$ (Fig. 1b).

This means that the distribution function for the current density reaching the surface in the phase interval from π to ${}^3/_2\pi$ will be cosinusoidal if we consider only electrons emitted over the phase interval 0-1.35. The phase interval $\pi - {}^3/_2\pi$ is precisely the one during which the secondary electrons emitted by the opposite surface can leave it in the direction of the original wall, since they will have a constant velocity component in the direction away from the surface. Figure 2 shows the range of phases of the field (corresponding to the interval $a_1 - a_2$ in Fig. 1b) in which it is possible for an electron emitted at phase φ_1 to encounter the wall.

Form of the Distribution Functions of the Primary and Secondary Currents

Since particles with the same translational velocity are positioned uniformly in the direction of motion with density n_1, their current i_1 per unit wall area over the active phase interval $\pi < \varphi < {}^3/_2\pi$ will vary as

$$i_1 = \frac{F}{m\omega}(-\cos\varphi + \cos\varphi_1)\, n_1, \tag{18}$$

if $0 < \varphi_1 < 1.35$. Multiplying and dividing by $\cos \varphi_1$, we obtain

$$i_1 = \frac{(-\cos \varphi + \cos \varphi_1)}{\cos \varphi_1} I_1, \tag{19}$$

where $I_1 = (Fn_1/m\omega)\cos \varphi_1$ is the time averaged flux of particles arriving at unit surface area with a translational velocity corresponding to an emission phase of φ_1.

Since in the following we shall regard the surface on which the electrons fall as an emitting surface, it is more convenient to shift the zero time for the phase φ by π so that the active interval of phases during which electrons can leave the surface ranges from 0 to $\pi/2$ rather than from π to $^3/_2\pi$ (the dashed axis of Fig. 1). In this case Eq. (19) takes the form

$$i_1 = \frac{\cos \varphi + \cos \varphi_1}{\cos \varphi_1} I_1. \tag{20}$$

Equation (20) gives the distribution of the current of primary particles with respect to the phase of arrival φ. The distribution of the secondary current with respect to the phase of escape is obtained by multiplying this by σ in accordance with Eqs. (14) and (15):

$$i(\varphi) = \frac{W}{W_1} \frac{(\cos \varphi + \cos \varphi_1)^3}{\cos \varphi_1} I_1. \tag{21}$$

Here $W = F^2/2m\omega^2$.

From this it is clear than an arbitrary flux I_1 of primary particles emitted in phase φ_1 creates a flux of secondary electrons which is distributed in the phase of emission φ in proportion to the factor

$$\cos^3 \varphi + a \cos^2 \varphi + b \cos \varphi + c, \tag{22}$$

where $0 \leq a \leq 3$, $0 \leq b \leq 3$, $0 \leq c \leq 1$, and the coefficients a, b, and c depend on the phase of emission φ_1 of the primary flux of electrons.

Here and everywhere in the preceding, φ_1 is the phase of emission of the primary electrons at one surface and φ is the phase of arrival of the primary electrons and the phase of emission of the secondary electrons at the other surface.

It is easy to show that the total secondary flux, equal to the sum of the secondary fluxes i produced by the primary electron fluxes I_1 with different emission phases φ_1, will have a distribution with respect to the phase of emission similar to that in Eq. (22): $D_1(\cos^3 \varphi_1 + A_1 \cos^2 \varphi_1 + B_1 \cos \varphi_1 + C_1)$, where D_1 is a factor measured in units of particles $\cdot$ cm^{-2} $\cdot$ sec^{-1}, and A_1, B_1, and C_1 are coefficients which take on arbitrary values within the limits specified above for a, b, and c.

It now remains to find the numerical values of A_1, B_1, and C_1.

An Operator for the Secondary Distribution in Terms of a Known Primary Distribution

We replace I_1 in Eq. (21) with $(D_1/2\pi)(\cos^3 \varphi_1 + A_1 \cos^2 \varphi_1 + B_1 \cos \varphi_1 + C_1)\Delta\varphi_1$, the averaged flux of electrons emitted over a narrow range of phase $\Delta\varphi_1 \ll \pi/2$. We shall integrate the resulting expression with respect to φ_1 from 0 to 1.35 by making the transition from $\Delta\varphi$ to $d\varphi$. As a result we obtain the distribution of the secondary electrons with respect to the phases of emission, $D(\cos^3 \varphi + A \cos^2 \varphi + B \cos \varphi + C)$:

$$\frac{W}{2\pi W_1} \int_0^{1.35} \frac{(\cos \varphi + \cos \varphi_1)^3}{\cos \varphi_1} D_1 (\cos^3 \varphi_1 + A_1 \cos^2 \varphi_1 + B_1 \cos \varphi_1 + C_1)\, d\varphi_1 = D(\cos^3 \varphi + A \cos^2 \varphi + B \cos \varphi + C). \tag{23}$$

We use the notation

$$2\pi D W_1 / D_1 W = H. \tag{24}$$

Expanding the brackets under the integral sign and equating the coefficients of equal powers of $\cos \varphi$ on the right- and left-hand sides of the equation, we find after the integration that

$$\begin{aligned} &0.976A_1 + 1.36B_1 + 2.5C_1 + 0.785 = H, \\ &2.36A_1 + 2.93B_1 + 4.072C_1 = HA, \\ &2A_1 + 2.36B_1 + 2.93C_1 + 1.76 = HB, \\ &0.588A_1 + 0.666B_1 + 0.785C_1 + 0.534 = HC. \end{aligned} \tag{25}$$

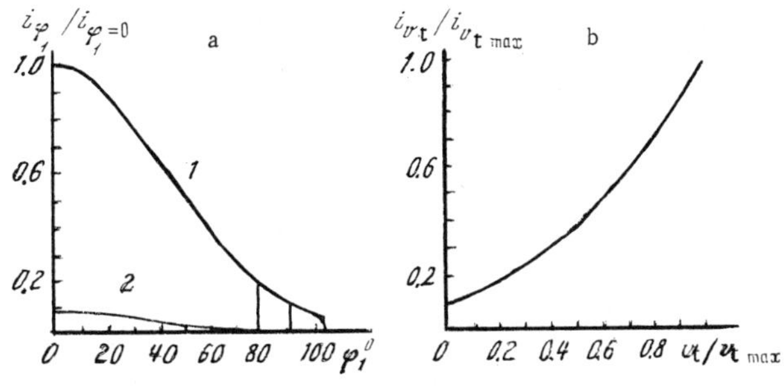

Fig. 3. The relative steady-state density of the flux of particles in a discharge as a function of the initial phase φ_1 (a) and of the relative translational velocity (b): (1) bulk distribution; (2) correction to the bulk distribution (scale magnified 5 times).

These equations give the relationship between the parameters of the primary flux (D_1, A_1, B_1, and C_1) and those of the secondary (D, A, B, and C).

The Steady-State Distribution Function and the Threshold Field

We seek a distribution function that does not vary when the flux interacts with a wall, that is, one such that $A_1 = A$, $B_1 = B$, $C_1 = C$. To do this we solve the system of equations (25) under the conditions $A_1 = A$, $B_1 = B$, $C_1 = C$.

Solution of Eq. (25) on a computer yielded the unique solution $A_0 = 2.18$, $B_0 = 1.78$, $C_0 = 0.51$, and $H_0 = 6.66$. This means that the main flux will be distributed with respect to the emission phase φ_1 as

$$i_1(\varphi) = D_1(\cos^3\varphi_1 + 2.18\cos^2\varphi_1 + 1.78\cos\varphi_1 + 0.51). \tag{26}$$

This equation gives the distribution of the flux with respect to the relative translational velocity (Fig. 3) since $\cos\varphi_1 = v_t m\omega / F$, where v_t is the translational velocity.

We now easily find the conditions under which the discharge will be self-sustaining. At the threshold power level, $D/D_1 = 1$. From Eq. (24) we find $W/W_1 = 2\pi/6.66$. Therefore, the threshold field required for production of a discharge can be found from the simple relation

$$W_{\text{thr}} = 0.94\, W_1. \tag{27}$$

Stability of the Distribution Function; Growth Rate of the Discharge

The function (26) is a solution of the problem stated at the beginning of this paper only if it is stable. We shall examine the stability of the distribution (26) with respect to small perturbations in A, B, and C from the values $A_0 = 2.18$, $B_0 = 1.78$, and $C_0 = 0.51$.

In the system (25) we replace B_1, A_1, C_1 by $A_0 + \Delta A_n$, $B_0 + \Delta B_n$, $C_0 + \Delta C_n$, respectively, where ΔA_n, ΔB_n, ΔC_n are very much smaller than A_0, B_0, and C_0, respectively, and seek the deviations ΔA_{n+1}, ΔB_{n+1}, and ΔC_{n+1} from A_0, B_0, and C_0 for the secondary function after the primary electrons have interacted with the wall. As a result, neglecting terms of second order of smallness relative to ΔA_n, ΔB_n, ΔC_n, we obtain

$$\begin{aligned}
\Delta A_{n+1} &= \Delta A_n \frac{k_{aa} - A_0 k_{ha}}{H_0} + \Delta B_n \frac{k_{ab} - A_0 k_{hb}}{H_0} + \Delta C_n \frac{k_{ac} - A_0 k_{hc}}{H_0}, \\
\Delta B_{n+1} &= \Delta A_n \frac{k_{ba} - B_0 k_{ha}}{H_0} + \Delta B_n \frac{k_{bb} - B_0 k_{hb}}{H_0} + \Delta C_n \frac{k_{bc} - B_0 k_{hc}}{H_0}, \\
\Delta C_{n+1} &= \Delta A_n \frac{k_{ca} - C_0 k_{ha}}{H_0} + \Delta B_n \frac{k_{cb} - C_0 k_{hb}}{H_0} + \Delta C_n \frac{k_{cc} - C_0 k_{hc}}{H_0}.
\end{aligned} \tag{28}$$

Here k_{aa}, k_{ab}, k_{ac} are the coefficients in system (25) and $H_0 = 6.66$. Substituting the numerical values, we have

$$\begin{aligned}
\Delta A_{n+1} &= 3.6\cdot 10^{-2}\Delta A_n - 4.5\cdot 10^{-3}\Delta B_n - 0.21\Delta C_n, \\
\Delta B_{n+1} &= 3.9\cdot 10^{-2}\Delta A_n - 9\cdot 10^{-3}\Delta B_n - 0.23\Delta C_n, \\
\Delta C_{n+1} &= 1.3\cdot 10^{-2}\Delta A_n - 5.1\cdot 10^{-3}\Delta B_n - 7.3\cdot 10^{-2}\Delta C_n.
\end{aligned} \tag{29}$$

If each successive deviation ΔA_{n+1}, ΔB_{n+1}, ΔC_{n+1} is smaller in absolute value than the preceding ΔA_n, ΔB_n, ΔC_n, then the function is stable. It is easy to confirm for the system (29) that always

$$|\Delta A_{n+1}| < |\Delta A_n|; \ |\Delta B_{n+1}| < |\Delta B_n|; \ |\Delta C_{n+1}| < |\Delta C_n|. \tag{30}$$

Therefore, the function (26) is stable with respect to small perturbations in A, B, and C.

We now estimate the growth rate of the discharge. Let τ be the time over which the bulk of the electrons is able to move from one surface to the other. From Eq. (24), when $H = H_0$, $D/D_1 = 1.06W/W_1$. From this it is easy to find how the discharge grows in time:

$$D_1 = D_0 \exp\left[\left(1.06\frac{W}{W_1} - 1\right)\frac{t}{\tau}\right], \tag{31}$$

where D_0 is the amplitude of the function in Eq. (26) at $t = 0$.

There is therefore one stable distribution [Eq. (26)] which can remain constant in time, and for this distribution we have found the value of the threshold field for appearance of a discharge. However, the question remains of whether there may exist a group of unstable functions with coefficients A, B, and C unequal to A_0, B_0, and C_0 which vary in time and have a threshold lower than that of Eq. (27). Clearly, there are no such functions, for otherwise (since it is always possible to isolate a small group of particles in the flux which are distributed arbitrarily in the phase of emission) we would observe a change in the form of the function (26) at the threshold field after one or more interactions with the wall. Since this is not observed, the very existense of a stable function is an indication that unstable functions with lower thresholds than Eq. (27) do not exist.

Equation (23) was integrated from 0 to 1.35, while the active part of the field is 0–1.57. This is because if $1.35 < \varphi_1 < 1.57$, then $a_1 > 0$ (see Fig. 1b, dashed axis) and particles with these emission phases do not yield the cosinusoidal distribution (20) over the entire active range $0 < \varphi < 1.57$. Besides this neglected phase interval, there is the interval $1.57 < \varphi_1 < 1.78$ (vertical dashed lines in Fig. 1a and b) which, although it is passive, does contribute to the discharge since the secondary electrons from this interval fall into the active interval $0 < \varphi < 1.57$ (dashed curves of Fig. 1a) after a time less than T. A study of the role of these "small" intervals under the assumption that the main flux of electrons is determined by Eq. (26) led to a refinement in the value of the threshold field (27) (see Fig. 3a, curve 2):

$$W_{\text{thr}} \approx 0.92W_1. \tag{32}$$

Since the maximum possible $W_c = 4W$, Eq. (32) implies that the above discussion is valid for materials in which σ is linear to values of at least 3.68. If the dependence of σ on W_c becomes a horizontal line for $\sigma < 3.68$, then a larger field than that implied by Eq. (32) is required for the onset of a uniform polyphase SED.

3. The Distribution Function of the Current and the Threshold for Onset of a Uniform Polyphase SED with Arbitrary Materials and Fields

In the above analysis it was assumed that the secondary emission coefficient rises linearly with the energy of the bombarding electrons. This approximation is valid for sufficiently small microwave fields. In general, however, there is obvious interest in solving the uniform polyphase SED threshold problem for functions $\sigma(W_c)$ as close to reality as possible (Fig. 4).

While preserving a general dependence of σ on W_c similar to that shown in Fig. 4 (dashed curve), various substances differ from one another in the maximum value σ_b and in the energy W_b at which $\sigma(W_c)$ bends, as well as in the slope of the curve at energies above W_b.

It is possible to greatly simplify the discussion if the real dependence is replaced by two straight lines as in Fig. 4. Then each material can be characterized by the three parameters σ_b, W_b, and W_0. With this simplification in the previous section it was possible to find an analytic function for the current distribution with respect to the phase of emission and for the threshold for onset of a discharge only when $W_c < W_b$, that is, when σ is directly proportional to the energy of the primary electrons.

In this section we shall also seek the current distributions with respect to the emission phase and the onset thresholds for the discharge, but for arbitrary values of W_c and for any substances whose secondary emissions coefficients have the form shown in Fig. 4. The search for the functions is done on a computer. As in Sec. 2, the search is based on the assumption that during the discharge there exists a stable distribution function of the discharge current with respect to the emission phase for arbitrary fields and dependences of

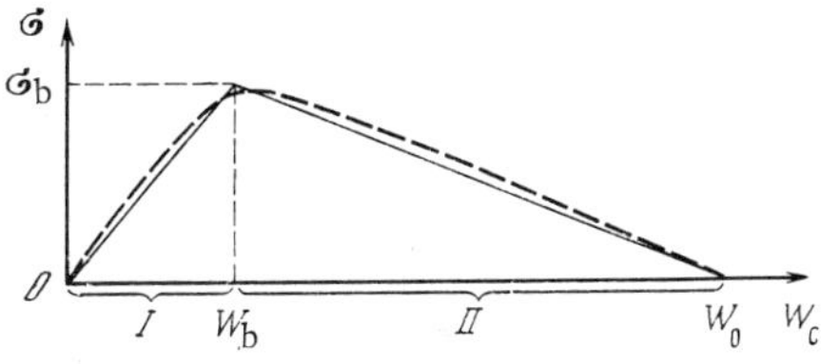

Fig. 4. A piecewise-linear approximation of the secondary emission characteristic.

σ on W_c. This function must change in form when the magnitude of the field or the secondary emission properties of the material are changed.

The energy of collision of an electron with the wall in the variable field is found using Eq. (15) and neglecting v_0:

$$W_c = W(\cos\varphi + \cos\varphi_1)^2, \tag{33}$$

where $W = F^2/2m\omega^2$, $\varphi = \omega t$, and $\varphi_1 = \omega t_1$.

The sign in front of $\cos\varphi$ [cf. Eq. (15)] is changed because the phase of the field, when it is measured relative to the surface on which the electron is incident after a long flight, is shifted by π relative to the field on the primary emission surface since the normals to the opposite walls are in opposite directions.

The following statement is very important: For given values of W_b, of W_0, and of the properties of the material which characterize its secondary emission, and for a given value of W characterizing the field, the form of the normalized distribution function of the current with respect to the phase of emission that is established in the discharge is independent of the magnitude of σ_b.

This phenomenon can be understood since, when the values of W_b, W_0, and W are kept constant, an increase or reduction in σ_b results in the same increase or reduction in the contribution from each phase of the field to the emission. We convinced ourselves of the correctness of this result with the aid of a computer calculation.

In addition, it is not the absolute values of W_b, W, and W_0 which are important, but their ratios. Thus, the two parameters $\beta = W_b/W$ and $\gamma = W_b/W_0$ fully define the form of the distribution function in a steady-state discharge and, therefore, specify all the possible stable distribution functions which can exist during a uniform polyphase discharge when microwave fields of arbitrary amplitude act on any solid material. Figure 5 shows selected graphs of the distribution function of the current with respect to the phase of emission for various values of the parameters β and γ. The process for obtaining these graphs with the aid of an operator which acts on an arbitrary primary distribution function to yield a secondary distribution function is described in the Appendix. This operator depends on β, γ, and σ_b.

If we take this operator with the values of β and γ of interest to us and with an arbitrary value of σ_b and act with it on a known, previously derived primary function which is stable for these values of β and γ (Fig. 5), then because the distribution function is independent of σ_b we obtain a secondary function which differs from the primary only by a constant factor which depends linearly on σ_b. After this it is easy to establish what value of σ_b must be put in the operator so that the constant factor in the secondary function should be equal to 1 (this

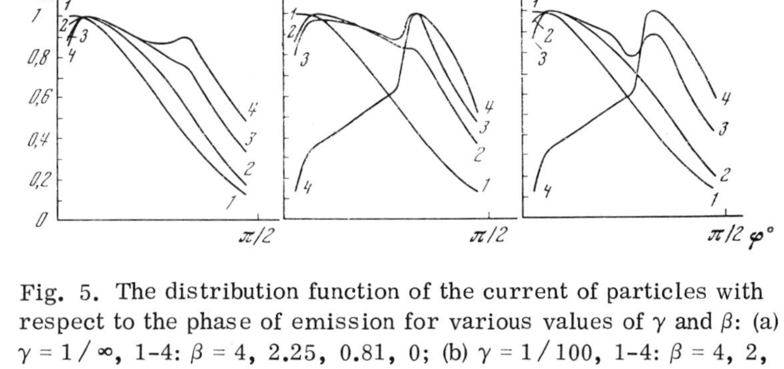

Fig. 5. The distribution function of the current of particles with respect to the phase of emission for various values of γ and β: (a) $\gamma = 1/\infty$, 1-4: $\beta = 4$, 2.25, 0.81, 0; (b) $\gamma = 1/100$, 1-4: $\beta = 4$, 2, 0.4, 0.03; (c) $\gamma = 1/25$, 1-4: $\beta = 4$, 0.8, 0.4, 0.12.

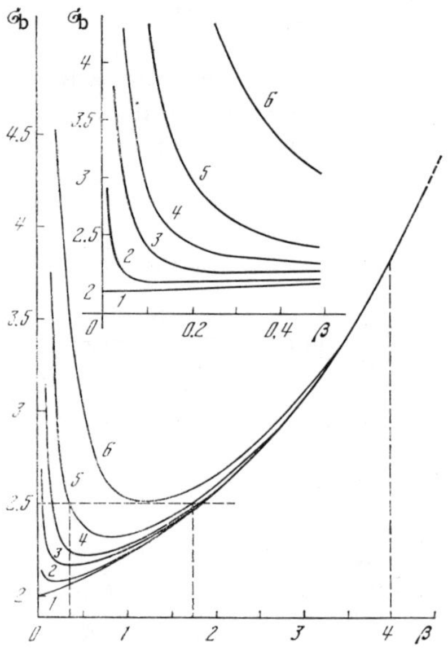

Fig. 6

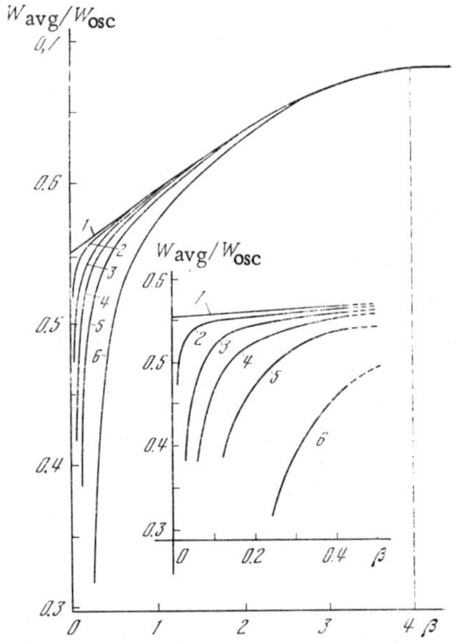

Fig. 7

Fig. 6. The dependence of the required σ_b for development of a uniform polyphase discharge on β for various values of γ: (1-6) $\gamma = 1/\infty$, 1/500, 1/100, 1/50, 1/25, 1/10, respectively. In the figure the beginning of the plot is shown with an enlarged scale.

Fig. 7. The dependence of the ratio of the translational energy of the electrons participating in the discharge to the oscillatory energy on β for various values of γ. The notation is the same as in Fig. 6.

means that the discharge will be self-sustaining, becoming neither stronger nor weaker). It was possible to obtain the threshold value of σ_b as a function of β for substances with different values of γ (Fig. 6). Using the curves of Fig. 6, one can determine the threshold field for a given material.

We now explain in detail how to use the graphs of Figs. 5 and 6. Consider a material with a known dependence of the secondary emission coefficient σ on the incident electron energy W_c (Fig. 4). Using the piecewise linear approximation, we have three characteristic parameters for the substance, σ_b, W_b, and W_0, together with $\gamma = W_b/W_0$. With Figs. 6 and 5 it is easy to find the field strengths at which a discharge can exist and the distribution function for a given field. As an example, we examine a material for which $\sigma_b = 2.5$ and $\gamma = 1/25$. Taking the point on the ordinate (see Fig. 6) corresponding to $\sigma_b = 2.5$ we draw a straight line from it parallel to the abscissa (the dashed line on the left-hand side of the figure) and find the points where this line intersects curve 5 which corresponds to our case of $\gamma = 1/25$. The coordinates of these intercepts on the β axis, 0.4 and 1.7, give the limiting values of β for which the discharge can occur. It is interesting that the region in which a discharge can occur is also bounded at high field strengths. We thus find that for this material a discharge can occur only at field strengths such that the condition $W_b/0.4 \geq W \geq W_b/1.7$ is satisfied. From this it is possible to find the minimum threshold value of W. From the graph (see Fig. 6) it follows that the greater σ_b is for a given value of γ, the larger the range of fields in which the discharge can occur. On the other hand, the larger the value of γ for the material, the larger the value of σ_b required for a discharge to occur. For $\beta \geq 4$ the secondary emission coefficient for all electrons which strike the wall is directly proportional to the energy; thus, at the point $\beta = 4$ all the curves merge into a single straight line. It is interesting to note (as is evident from Fig. 6) that no matter how large the field is, a uniform polyphase discharge cannot develop if $\sigma_b < 1.96$.

Using Fig. 5 we find the normalized distribution function of the discharge current with respect to the emission phase for various values of W lying within the range where the discharge exists. From the figure it is clear that, if the energies of all particles hitting the wall are below W_b (and this will be the case when $\beta = W_b/W \geq 4$), then the steady-state function coincides with that obtained analytically in Sec. 2 (where the coefficient σ is directly proportional to the collision energy). As the field is increased, the form of the function changes and its maximum is shifted from the point $\varphi_1 = 0$ toward larger values of the emission phase φ_1. This signifies, according to Eq. (33), a reduction in the energy of translational motion of the particles. Figure 7 shows the field dependence of the ratio of the average translational energy per electron in the discharge, W_{avg}, to the energy of oscillatory motion W, where W_{avg} is evaluated using the formula

$$W_{avg} = \int_0^{\pi/2} W \cos^2 \varphi_1 \cdot i_1(\varphi_1)\, d\varphi_1 \left[\int_0^{\pi/2} i_1(\varphi)\, d\varphi \right]^{-1}.$$

The curves illustrate the reduction in translational energy, averaged over all electrons emitted in the range 0-90°, as the field is increased.

4. The Distribution Function of the Electrons and the Threshold for Appearance of a Uniform Polyphase SED when E ∥ S

The situation examined in the preceding sections, where the electric vector of the microwave field is orthogonal to the sample surface, is of interest during irradiation of both metallic and dielectric surfaces by microwaves.

The range of problems in which the phenomenon of the uniform polyphase SED is important, however, also includes the case of parallel field orientation (E ∥ S). This type of interaction can occur when waves are incident on dielectric samples, in particular at the output windows of high-power microwave oscillator tubes or at the transparent inlet windows through which the radiation is fed into magnetic confinement devices (in thermonuclear research).

The development of a uniform polyphase SED with E ∥ S can be treated by means of the method suggested in Secs. 2 and 3. It is then necessary to establish the form of the distribution function of the electrons with respect to the phase of emission.

Let E ∥ S at the surface. The field acts on an electron with force $F \sin \omega t$ and the secondary electrons have a Maxwellian distribution for the component of velocity perpendicular to the surface. When the restoring force is small, if the time the electron spends in flight obeys $t_f \gg T$, then the arrival of an electron at the surface is equally probable for any phase. A current i_1 emitted in the range $\Delta\varphi_1$ of the phase (of the field) yields an average density $I_1 = i_1 \Delta\varphi_1 / 2\pi$ (where i_1 is the distribution of the primary current with respect to the phase of emission φ_1) upon collision with the surface because of the equal probability of arrival of an electron in an arbitrary phase. We assume, as in Sec. 2, that $\sigma = W_c / W_1$. The electrons emitted over the interval $\Delta\varphi_1$ "give birth to" a secondary flux

$$i(\varphi) = \frac{i_1 \Delta\varphi_1}{2\pi} \frac{W}{W_1} (\cos \varphi_1 + \cos \varphi)^2. \tag{34}$$

This equation means that an arbitrary primary flux i_1 emitted at phase φ_1 will, when it returns to the "mother" surface, initiate a secondary flux with a distribution in the phase of emission proportional to the factor $\cos^2 \varphi + a \cos \varphi + b$, where $2 \geq a \geq -2$ and $1 \geq b \geq 0$.

It is easily shown that the total secondary flux, which is made up of the secondary fluxes produced by primary electron currents with different emission phases φ_1, will also have a phase distribution of the form $D(\cos^2 \varphi + A \cos \varphi + B)$ (A and B can vary over the same limits as a and b). Since it is known that a flux of primary particles with a distribution in the emission phase φ_1 of the form $i_1 = D(\cos^2 \varphi_1 + A_1 \cos \varphi_1 + B_1)$ is incident on the surface, it is possible to derive the total distribution of the secondary current with respect to the phase of emission by integrating Eq. (34) over φ_1 from 0 to 2π:

$$\int_0^{2\pi} \frac{D_1 W}{2\pi W_1} (\cos^2 \varphi_1 + A_1 \cos \varphi_1 + B_1)(\cos^2 \varphi_1 + 2 \cos \varphi_1 \cos \varphi + \cos^2 \varphi)\, d\varphi_1 = D(\cos^2 \varphi + A \cos \varphi + B). \tag{35}$$

Using the notation $2\pi D W_1 / D_1 W = H$, as before, and equating the coefficients of equal powers of $\cos \varphi$ on the left- and right-hand sides of Eq. (35), we obtain the system of equations

$$\int_0^{2\pi} (\cos^2 \varphi_1 + A_1 \cos \varphi_1 + B_1)\, d\varphi_1 = H,$$

$$2 \int_0^{2\pi} (\cos^3 \varphi_1 + A_1 \cos^2 \varphi_1 + B_1 \cos \varphi_1)\, d\varphi_1 = HA,$$

$$\int_0^{2\pi} (\cos^4 \varphi_1 + A_1 \cos^3 \varphi_1 + B_1 \cos^2 \varphi_1) \, d\varphi_1 = HB. \tag{36}$$

Integrating Eq. (36) we obtain a system of equations which relates the parameters of the primary current A_1, B_1, and D_1 to the secondary A, B, and D:

$$\begin{aligned} \pi + 2\pi B_1 &= H, \\ 2\pi A_1 &= HA, \\ 0.75\pi + \pi B_1 &= HB. \end{aligned} \tag{37}$$

Rewriting Eq. (37) we obtain

$$D = (0.5 + B_1) D_1 \frac{W}{W_1}, \tag{38a}$$

$$A = \frac{A_1}{0.5 + B_1}, \tag{38b}$$

$$B = \frac{0.75 + B_1}{1 + 2B_1}. \tag{38c}$$

We now analyze this system (38). On comparing the three equations, we see that the quantity D which characterizes the secondary current is a function of the two primary current parameters D_1 and B_1. The same may be said of the parameter A, which depends on A_1 and B_1. At the same time, B depends only on B_1. It is clear from Eq. (38c) that after several interactions $B = B_1 = B_0 \approx 0.613$. It is then easy to confirm that, for an arbitrary deviation of B_1 from B_0 in the primary function, the deviation of B from B_0 in the secondary function will be less than that of B_1 from B_0, so that the value $B = B_0$ is stable.

Substituting $B_1 = 0.613$ in Eq. (38b), we obtain

$$A \approx 0.9 A_1. \tag{39}$$

Equation (39) means that each succeeding distribution function will have a parameter A smaller than the parameter A_1 of the previous distribution function. Evidently, the final result is a stable distribution function with $A = A_0 = 0$.

It is thus clear that the distribution with respect to the phase of emission that is established in the final stage of the distribution will have the form

$$D\,(\cos^2 \varphi_1 + 0.613). \tag{40}$$

Since $D/D_1 = 1$ as the threshold conditions are approached, it is easy to obtain the threshold field from Eq. (38a):

$$W_{\text{thr}} = 0.88 W_1. \tag{41}$$

With the above analysis we have therefore determined the distribution function of the discharge current with respect to the phase of emission [Eq. (40)] and the threshold field [Eq. (41)].

It is interesting to compare the criterion (41) obtained from a rigorous solution with that derived in [16]. Since $W_{\text{thr}} = e^2 E_{\text{thr}}^2 / 2m\omega^2$, from Eq. (41) we obtain

$$E_{\text{thr}} = 0.94 \omega \sqrt{2U_1/\eta}. \tag{42}$$

Here $\eta = e/m$ and $U_1 = W_1/e$ is the first critical potential (volts).

Therefore, our value of E_{thr} is close to the qualitative criterion introduced in [16]:

$$E_{\text{thr}} \simeq \omega \sqrt{2U_1/\eta}.$$

5. Threshold for Appearance of a Uniform Polyphase SED for Arbitrary Angles between the E Vector of the Microwave Field and the Surface

For examining the case in which the electric vector of the alternating field is at an angle β to the normal of the surface, we may use the methods and results obtained for $E \perp S$ ($\beta = 0$).

It is easy to see that when there is an angle β between the normal to the surface and E, the motion of an emitted electron can be described by two orthogonal motions: one in a direction perpendicular to the surface with a velocity $\dot{x} = (F/m\omega)(-\cos\varphi + \cos\varphi_1)\cos\beta$ and the other along the surface with velocity $\dot{y} = (F/m\omega) \times (-\cos\varphi_1 + \cos\varphi_1)\sin\beta$.

Evidently, in this case it is also possible to write down an expression for the current of particles emitted in phase φ_1 in the form of Eq. (20), except that here

$$I_1 = \frac{Fn_1}{m\omega}\cos\varphi_1 \cdot \cos\beta. \tag{43}$$

Since the motion along the surface has no effect on the probability of incidence of an electron at a given phase of the field, all our conclusions about the working range of phases for the primary and secondary currents remain in force. As for the number of secondary electrons knocked out by the primaries, this depends on the total velocity of an electron and the magnitude of W_1 (the energy at which $\sigma = 1$). This energy depends on the angle of incidence β of the primary electrons. The emission phase distribution of the secondary current [Eq. (21)] has the following form when this dependence is taken into account:

$$i(\varphi) = \frac{W}{W_1(\beta)}\,\frac{(\cos\varphi + \cos\varphi_1)^3}{\cos\varphi_1}\,I_1. \tag{44}$$

An analysis of Eq. (23) leads to the conclusion that the nature of the coupling between the primary and secondary fluxes does not change when β is taken into account; that is, Eq. (23) retains its form. The only new feature is the need to include the dependence of W_1 on β; that is, in Eq. (23) W_1 must be replaced by $W_1(\beta)$.

Finally, the threshold field at which the current to the surface begins to rise can be found from the equation

$$W_{\text{thr}} = 0.92 W_1(\beta). \tag{45}$$

Naturally, all the above discussion is valid only so long as the velocity acquired by an electron from the microwave field in the direction perpendicular to the surface is much greater than the "thermal" velocity of the electrons; i.e.,

$$\frac{F}{m\omega}\cos\beta \gg \sqrt{\frac{kT_e}{m}}, \tag{46}$$

where k is Boltzmann's constant and $T_e \sim 3.3 \cdot 10^4$°K.

When Eq. (46) is not satisfied electrons emitted outside the working range of phases established in Sec. 2 begin to play a significant role. Electrons appear in the discharge which have been emitted at phases of the field such that the normal component of F is directed toward the surface. In the limiting case where the normal component of the velocity acquired from the field is much less than the "thermal" velocity, the latter is dominant in the motion of the electrons perpendicular to the surface. It is then possible to use the criterion (44) for the case E∥S obtained in Sec. 4.

We shall consider the threshold for a uniform polyphase SED under the condition that the current from outside does not lie in the region where the microwave field interacts with the material. Condition (45) means that if the current enters the interaction region, then when the criterion (45) is satisfied, this current will grow along the surface. When $\beta = 0$, fulfillment of the condition for a growth in the current implies that a single random electron entering the microwave interaction region will cause an avalanchelike growth in the number of electrons in this region. When $\beta \neq 0$, there is a "removal" velocity $\dot{y} = (F/m\omega)(-\cos\varphi + \cos\varphi_1)\sin\beta$ which causes the secondary electrons to strike the surface a distance $t_f(F/m\omega)\cos\varphi_1 \cdot \sin\beta$ from the emission point in the direction of the "removal" velocity. This effect may cause the region where the electrons exist to move in the y direction at the average "removal" velocity and after a certain time the interaction region will be cleared of electrons. Although criterion (45) will be satisfied, a self-sustaining discharge may not develop. Thus, when $\beta \neq 0$ and $\beta \neq 90°$ an additional criterion (besides the one that the current falling into the interaction region must be amplified) is necessary for determining the conditions under which the discharge can be "held" onto a surface with oblique E.

When $\beta = 0$ two opposing currents in the y direction caused by the Maxwellian distribution of the "thermal" velocities may be distinguished. The flux in the +y direction aids expansion of the discharge to the right of the region where it first appears and that in the −y, to the left. When E is inclined with respect to the surface the resulting "removal" velocity $v_0 \simeq (F/m\omega)\sin\beta$ causes a reduction in one of the two opposing currents. If $\langle\sigma\rangle$, the secondary emission coefficient averaged over all phases of the field, can compensate the reduced flux due to v_0 for a given β, then it is evident that the discharge can be held on the surface. Otherwise all the electrons would "run off." Thus, the criterion for a self-sustaining discharge is

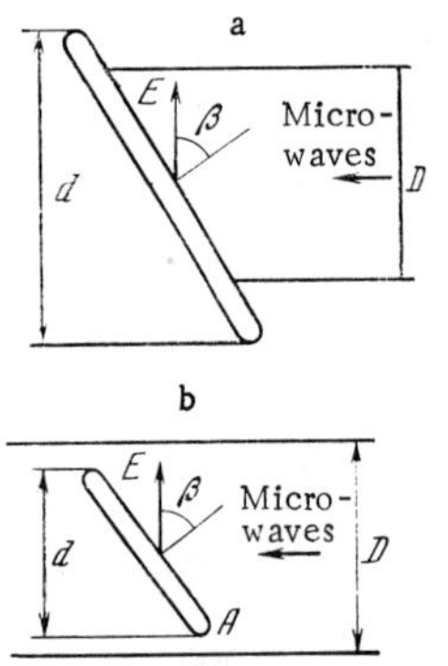

Fig. 8

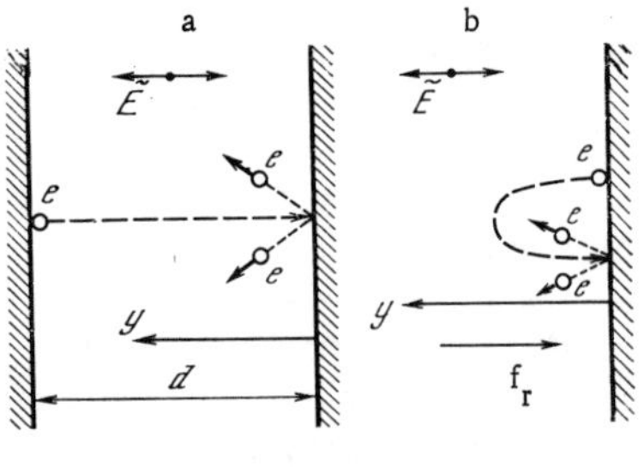

Fig. 9

Fig. 8. An example illustrating the difference in the conditions for development of a discharge: (a) D < d; (b) D > d.

Fig. 9. The geometric conditions for the formation of two-sided (a) and one-sided (b) discharges.

$$\frac{I_{v_0}}{I_0}\langle\sigma\rangle \geqslant 1. \tag{47}$$

Here I_0 is the flux of particles in the y direction when there is no "removal" velocity and I_{v_0} is the flux of particles including the "removal" factor.

Using the Maxwellian distribution of the electron velocity component along the y axis

$$f(v_y)=\left(\frac{m}{2\pi kT}\right)^{1/2}\exp\left(-\frac{mv_y^2}{2kT}\right),$$

we obtain

$$\frac{I_{v_0}}{I_0}=\frac{\int_0^\infty f(v_y)(v_y-v_0)\,dv_y}{\int_0^\infty f(v_y)\,v_y dv_y}. \tag{48}$$

Assuming that $mv_0^2/2 \gg kT$, after integration of Eq. (48) we have

$$\frac{I_{v_0}}{I_0}=\left(1+\frac{1.45v_0}{\sqrt{kT/m}}\right)\exp\left(-\frac{mv_0^2}{2kT}\right). \tag{49}$$

Since $H = H_0 = 6.66$, we obtain the value $\langle\sigma\rangle = D/D_1 = 1.06W/W_1(\beta)$ from Eq. (24). In conclusion, condition (47) can be rewritten in the form

$$1.06\,\frac{W}{W_1(\beta)}\left[1+\frac{1.45\,(F/m\omega)\sin\beta}{\sqrt{kT/m}}\right]\exp\left[-\frac{m}{2kT}\left(\frac{F}{m\omega}\sin\beta\right)^2\right]\geqslant 1. \tag{50}$$

Therefore, for E inclined to the surface there are two threshold criteria. One [Eq. (45)] determines the threshold above which amplification of an outside current will occur, but a self-sustaining discharge cannot exist. Exceeding the second criterion [Eq. (50)] means that the conditions for a self-sustaining discharge are satisfied. When $\beta = 0$ and $E \perp S$, Eq. (50) also degenerates to Eq. (45). These conclusions can be demonstrated by two examples: let the cross section of the region containing the microwave field, D, obey D < d, where d is the cross section of an object in free space (Fig. 8a). In this case criterion (50) must be used if there is no electron emitter on the object itself. If, for the same field level, D > d and somewhere on the edge, even at a single point A (Fig. 8b), the more "rigid" condition (50) is satisfied, then the entire plane will be engulfed in a discharge, even if on the plane only the "weaker" condition (45) were satisfied and (50) were not.

As a conclusion to this section it is interesting to examine the way the threshold field in (45) and (50) changes with β using the known dependence of σ on the angle of incidence of the primary electrons. In [17] the dependence

$$\sigma = Be^{-\gamma\cos\beta} \tag{51}$$

was derived theoretically and verified experimentally. Using Eq. (45) we may write for the ratios of the threshold fields that

$$W_{thr}(\beta = 0)/W_{thr}(\beta) = W_1(\beta = 0)/W_1(\beta).$$

Since, under the assumption that σ is directly proportional to the energy, $W_1(\beta = 0)/W_1(\beta) = \sigma(\beta)/\sigma(\beta = 0)$, we have $W_{thr}(\beta = 0)/W_{thr}(\beta) = \exp[-\gamma(\cos\beta - 1)]$. Using the relation $\sigma(\beta = 75°\text{-}80°)/\sigma(\beta = 0) = 1.5\text{-}2 \simeq 1.75$ [17], we find $\gamma = 0.56$.

Thus, finally,

$$W_{thr}(\beta)/W_{thr}(\beta = 0) = \exp[0.56(\cos\beta - 1)]. \tag{52}$$

6. Theoretical Relationship between the Secondary-Emission Properties of a Substance and the Threshold Restoring Force

For breakdown in a two-sided discharge it is sufficient that the threshold value of the microwave electric field be exceeded. There is another requirement in the case of a one-sided discharge: a sufficiently large restoring force f_r must be present.

The requirement that a threshold restoring force be exceeded can be explained from a physical standpoint as follows. When f_r is absent all secondary electrons leave the target and the conditions for electron multiplication and avalanche formation are absent. When $f_r \neq 0$ part of the secondary electrons which leave the isolated electrode will return to the emitting surface (the second generation of particles). By increasing f_r it is possible to reach a condition such that the number of electrons in each successive generation exceeds the number of electrons in the preceding generation. An avalanche develops when the condition $\langle\sigma\rangle \geq 1$ is satisfied. Here

$$\langle\sigma\rangle \equiv \int_0^{W_{c\,max}} \sigma(W_c) F(W_c)\, dW_c, \tag{53}$$

where $F(W_c)$ is the normalized energy distribution function of the electrons bombarding the plate and $W_{c\,max}$ is the maximum energy of incidence.

The problem of the development of a discharge on a single surface can be solved for given geometric conditions (Fig. 9b) within the framework of the following rather plausible approximations:

(a) it is assumed that the secondary emission coefficient of the surface depends on the impact energy in accordance with Eq. (14), and

(b) the restoring force f_r is related to the existence of a positive potential U_0 on the plate which interacts with the microwave field. The force f_r has no component parallel to the surface of the plate and its magnitude is independent of the y coordinate.

It follows from (a) that the determination of the averaged secondary electron emission coefficient reduces to finding the energy W_c with which the electrons strike the plate. This energy is found by solving the equation of motion of an electron:

$$m\frac{d^2y}{dt^2} = eE_m \sin(\omega t_f + \varphi) - f_r. \tag{54}$$

Under the above assumptions

$$W_c = W[\cos(\varphi_1 + \varphi) + \cos\varphi]^2, \tag{55}$$

where $\varphi_1 = \omega t_f$, φ is the phase of emission, and $W = e^2E_m^2/2m\omega^2$ is the maximum energy of oscillatory motion of the electrons in the microwave field.

The energy distribution of the electrons bombarding the plate can be written in the form

$$F(W_c) = F_1(\varphi_1)F_2(\varphi). \tag{56}$$

With Eqs. (54)-(56), Eq. (53) now takes the form

$$\langle\sigma\rangle = \frac{W}{W_1}\int_{\varphi_0}^{\pi/2}\int_{-\varphi}^{\pi/2-\varphi} [\cos(\varphi_1 + \varphi) + \cos\varphi]^2 F_1(\varphi_1) F_2(\varphi)\, d\varphi\, d\varphi_1. \tag{57}$$

The limits of integration follow from an analysis of the motion of electrons in the microwave field. The interval of phases of the field over which secondary electrons can escape the target satisfies the condition

$$0 < \varphi_1 + \varphi < \pi/2. \tag{58}$$

Meanwhile, only those electrons whose drift velocity is sufficiently small that the force f_r can return them to the emitting surface can participate in the discharge. Consequently, the range of phases of emission for the electrons that form the avalanche is

$$\varphi_0 \leqslant \varphi \leqslant \pi/2, \tag{59}$$

where φ_0 is found from the equality of the translational energy of an electron to the work required to overcome the potential U_0,

$$W \cos \varphi_0 = eU_0. \tag{60}$$

When part of the most energetic electrons is unable to return to the plate it is rather difficult to find the distribution of the electrons with respect to φ and φ_1. Thus, it is necessary to limit ourselves to a qualitative examination of the process. For this we shall assume that electrons bombarding the plate are distributed uniformly in the phases φ and φ_1 over the intervals determined by inequalities (58) and (59). Then the distribution function (56) is constant; i.e.,

$$F(W_c) \equiv C. \tag{61}$$

Finally, after integration of Eq. (57) the criterion for development of a one-sided uniform polyphase SED takes the form

$$C \frac{W}{W_1} \Phi(\alpha) \geqslant 1, \tag{62}$$

where $\alpha \equiv eU_0/W$, $C = \text{const}$, and

$$\Phi(\alpha) \equiv \frac{\pi}{4}\left(\frac{\pi}{2} - \arccos\sqrt{\alpha}\right) + 2\left(1 - \sqrt{1-\alpha}\right) + \frac{\pi}{2}\left[\frac{\pi}{4} - \frac{\arccos\sqrt{\alpha}}{2} - \frac{\sqrt{\alpha(1-\alpha)}}{4}\right]. \tag{63}$$

The function (63) is plotted in Fig. 10. Assuming that $\Phi(\alpha)$ is linear, we find that the threshold value of the plate potential beyond which breakdown can be expected to occur is

$$eU_{0\,\text{thr}} = W_1/C_1. \tag{64}$$

From this equation it follows that the threshold potential is independent of the electric field of the microwaves (when, of course, the threshold with respect to E_m has already been exceeded) and rises linearly with the first critical potential W_1/e.

7. Determination of the Floating Potential of a Target

Since one possible mechanism for the appearance of a restoring force is electrostatic charging of the target, it is of obvious interest to calculate the value of the so-called floating potential U_f of the isolated plate. We define U_f from the condition that the total current to the plate be zero. This latest requirement reduces to a need that the following equation be satisfied:

$$\frac{4}{\pi^2}\frac{W}{W_1}\int_0^{\varphi_0}\int_{-\varphi}^{\pi/2-\varphi} \left[\cos^2(\varphi_1+\varphi) + 2\cos(\varphi_1+\varphi)\cdot\cos\varphi + \cos^2\varphi\right] d\varphi_1\, d\varphi = 1. \tag{65}$$

This equation was obtained analogously to Eq. (57) but the basic consideration here is that the floating potential of the target, and it rises from zero, is stabilized when the number of particles coming from outside equals the number of particles leaving the target and able to overcome the floating potential U_f defined by Eq. (60) as U_0. Since the bulk of the particles arrives at the surface and is emitted within the phase interval 0–$\pi/2$, it is natural to set $F_1(\varphi) = F_2(\varphi) = 2/\pi$ in a rough approximation.

After integrating Eq. (65) we obtain

$$\Phi_1(\alpha_1) = W_1/W, \tag{66}$$

where $\alpha_1 = eU_f/W$ and $\Phi_1(\alpha_1)$ is a function of the form (Fig. 11)

$$\Phi_1(\alpha_1) \equiv \frac{\arccos\sqrt{\alpha_1}}{\pi} + \frac{8}{\pi^2}\sqrt{1-\alpha_1} + \frac{1}{\pi}\left[\arccos\sqrt{\alpha_1} + \frac{1}{2}\sqrt{\alpha_1(1-\alpha_1)}\right]. \tag{67}$$

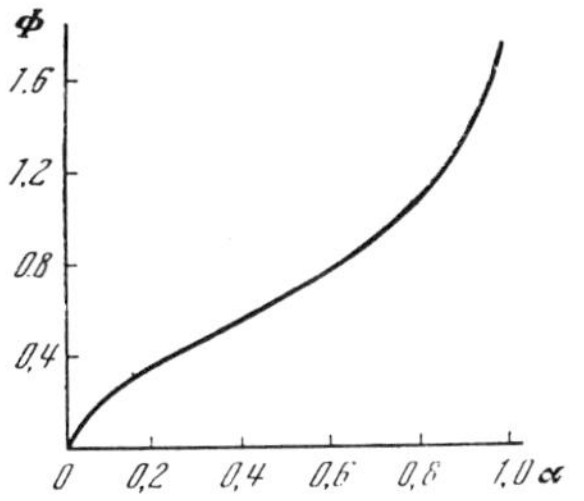

Fig. 10

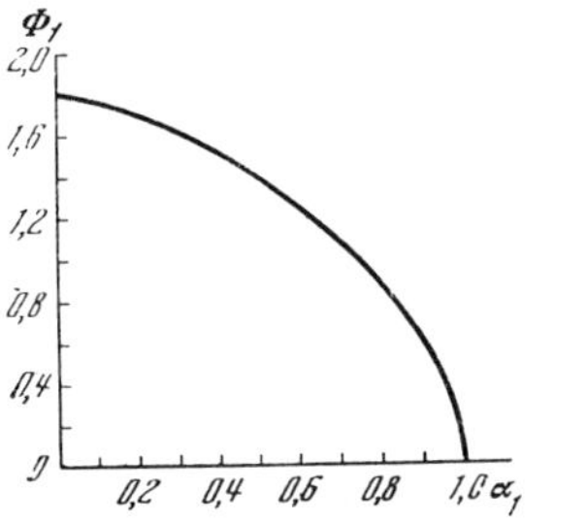

Fig. 11

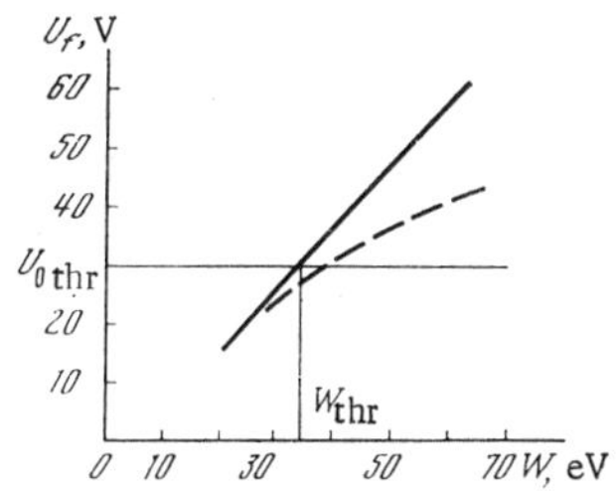

Fig. 12

Fig. 10. The dependence of Φ on α.

Fig. 11. The dependence of Φ_1 on α.

Fig. 12. The theoretical dependence of the floating potential V_f on the microwave field. The smooth curve is for $\sigma = W_c / W_1$; the dashed curve is the dependence when $\sigma(W_c)$ has the form shown in Fig. 4.

Using Eq. (66) and the graph of Fig. 11, we can find the value of the floating potential U_f for given W and W_1. An example of such a calculation for a substance with $W_1 = 22$ V is shown in Fig. 12.

If the floating potential reaches $U_{0\,thr}$ a self-sustaining discharge develops on the target. Evidently the floating potential begins to rise from zero only when the microwave field is at a level above threshold. The greater the microwave field, the greater U_f is.

Graphical calculations of the type shown in Fig. 12 could be used for determining the threshold fields for materials with secondary emission characteristics in the form of Eq. (14) and in the form shown in Fig. 4 (the dashed curve of Fig. 12). The methods described in Secs. 2 and 3 could be used to calculate these curves; however, these calculations would be extremely tedious.

The material in this section should thus be viewed only as a qualitative indication that the power threshold at an isolated target will probably be higher than at a target with an externally applied potential.

If an isolated body subjected to the action of a microwave field is electrically neutral, then it is necessary to determine the time τ for the floating potential to reach $U_{0\,thr}$. Evidently, the delay time τ between the beginning of the microwave interaction and the moment a uniform polyphase SED appears can be estimated using the formula

$$\tau \simeq \frac{U_{0\,\mathrm{thr}} C_s}{i_{\mathrm{inc}}(\langle\sigma\rangle - 1)}, \quad (68)$$

where C_s is the capacitance per unit surface area and i_{inc} is the density of the current arriving from outside.

8. The Restoring Force Created by a Gradient in the Microwave Field

A restoring force can also appear without electrostatic charging if a Miller force of sufficient strength is present. Thus, if a plane wave of amplitude E_{in} is incident from a vacuum on a dielectric surface with a dielectric constant ε, then according to [18] the field on the surface is equal to E_{in} and the amplitude of the reflected wave, E_{refl}, is found from

$$E_{\mathrm{refl}} = E_{\mathrm{in}}\left(\frac{1-\sqrt{\varepsilon}}{1+\sqrt{\varepsilon}}\right).$$

Evidently, the resulting standing wave will have an electric field at its antinode of

$$E_{\mathrm{cr}} = E_{\mathrm{in}}\left(1 + \left|\frac{1-\sqrt{\varepsilon}}{1+\sqrt{\varepsilon}}\right|\right).$$

The restoring force will be sufficient to produce a discharge if the difference between the quasipotential at the surface and that in the antinode is greater than the component of the electron energy (~4 eV) normal to the surface; i.e., when

$$\frac{e^2 E_{\mathrm{in}}^2}{4m\omega^2}\left(2\left|\frac{1-\sqrt{\varepsilon}}{1+\sqrt{\varepsilon}}\right| + \left|\frac{1-\sqrt{\varepsilon}}{1+\sqrt{\varepsilon}}\right|^2\right) > 4\ \mathrm{eV}. \quad (69)$$

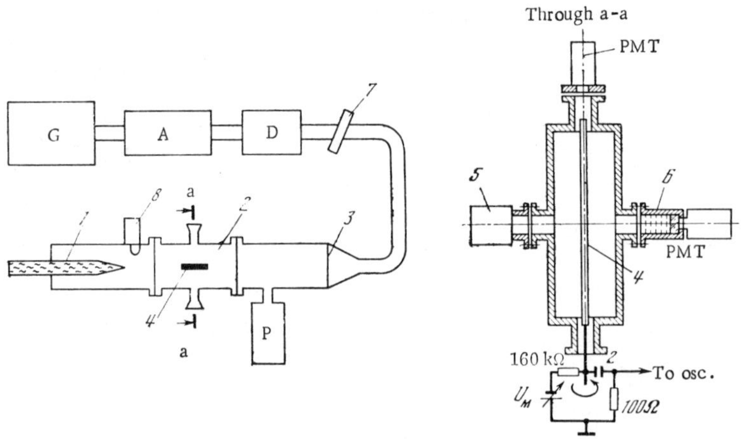

Fig. 13. A diagram of the apparatus.

CHAPTER II

EXPERIMENTAL DETERMINATION OF THE THRESHOLD FOR A UNIFORM POLYPHASE SED

9. Experimental Apparatus

The uniform polyphase secondary-emission discharge model analyzed in the preceding sections has not been investigated experimentally (nor has the nonresonant discharge model adopted in [14]). It is necessary to find whether a microwave discharge of the uniform polyphase SED type actually exists and, if it exists, whether it has the properties which follow from our theoretical analysis.

To study vacuum secondary-emission breakdown in cases where the rigorous conditions for a secondary-emission resonance discharge are not satisfied, we have built the experimental apparatus illustrated in Fig. 13. The generator G produces high-frequency pulses lasting 8 μsec at a wavelength of ~10 cm. The output power can be varied over the range from 0 to 1 MW. The microwave power passes through a ferrite attenuator A into a power divider D. With the aid of the divider the microwaves can be directed either into a calorimeter power meter or through a directional coupler 7, with which the incident and reflected power can be measured, and on to an evacuated waveguide 2 which contains the test target 4. Beyond the target, the wave reaches a matched load 1. The matched load was sometimes replaced by a short-circuiting piston with which a standing wave regime could be created in the waveguide. The power which passed by the target could be measured by means of a detector head with a magnetic pickup loop 8.

The waveguide section included a glass cone 3 for isolating the evacuated portion of the waveguide circuit. The evacuated portion was pumped by an oil-free electrical discharge titanium pump P (NÉM-1-S) to a residual gas pressure of 10^{-6}-$3 \cdot 10^{-7}$ torr.

During operation the stainless steel evacuated portion of the waveguide could be heated to a temperature of 300-400°C. This part of the waveguide had a cross section corresponding to that of a standard rectangular 72 × 34 mm waveguide in which the lowest H_{10} mode is excited.

With an H_{10} wave the electromagnetic field is invariant along the narrow wall of the waveguide. Along the wide wall the field varies sinusoidally. If we choose coordinates with the z axis along the axis of the waveguide, the y axis along the wide wall, and the x axis along the narrow wall, then the components of the microwave field in the waveguide (traveling wave regime) will have the form

$$E_y = 0, \quad E_z = 0, \quad E_x = E_m \sin \frac{\pi}{a} y \cdot \sin(\omega t - kz),$$

$$H_x = 0, \quad H_z = H_{mz} \cos \frac{\pi}{a} y \cdot \cos(\omega t - kz), \quad H_y = H_{my} \sin \frac{\pi}{a} y \cdot \sin(\omega t - kz).$$

The target is a metallic or dielectric plate of length ~10 cm, width 2 cm, and thickness 1-2 mm. The ends of the target enter ports with diameters of ~3 cm on the narrow walls of the waveguide. The lower end of the target is attached to a cylindrical metal rod whose end passes through a vacuum seal to the outside. By

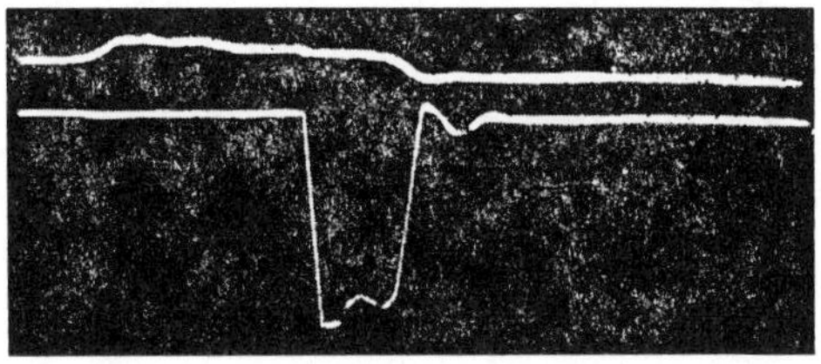

Fig. 14

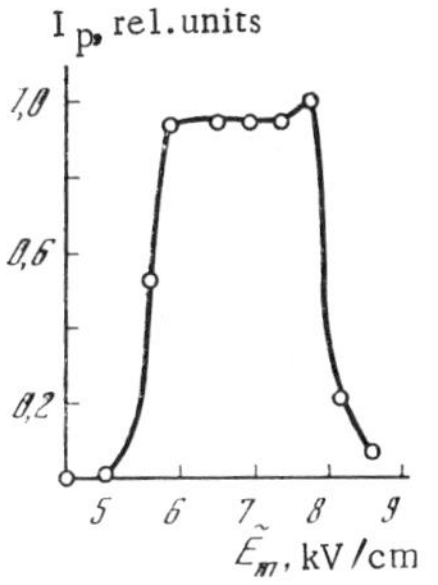

Fig. 15

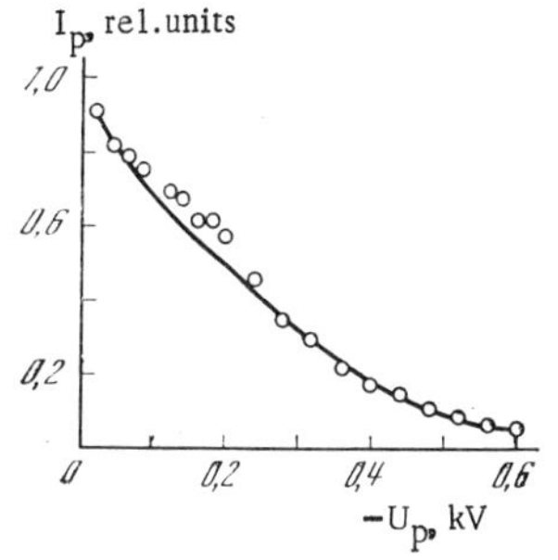

Fig. 16

Fig. 14. Oscilloscope traces of the envelope of the microwave pulse (top) and of the signal from the electrostatic probe for an empty waveguide (bottom).

Fig. 15. The dependence of the current from an empty waveguide on the magnitude of the alternating electric field.

Fig. 16. A current retardation curve for the empty waveguide.

rotating the outside end of the rod it is possible to turn the target at an arbitrary angle to the electric field lines of the wave. The target and supporting rod are galvanically isolated from the walls of the waveguide.

The axis of the target is located at a cross section with four ports. In the middle of the wide wall facing the target there is a multigrid probe 6 to detect the electron current from the target. On the opposite side there is an electron gun 5 which was needed for studying the secondary-emission characteristics of the targets (see the Appendix). A photomultiplier tube (PMT) was attached to the remaining port on the narrow wall of the waveguide for recording the luminosity of the sample in the microwave field when dielectric targets were used.

10. Study of a Two-Sided Polyphase Discharge

Prior to placing the test samples in the waveguide, it was appropriate to examine the processes which occur in the empty waveguide when the microwave field is applied, so that the effects caused by the interaction of the waves with the walls of the waveguide could be taken into account during studies of the interaction of the microwaves with samples.

An investigation of phenomena in the empty waveguide is also of independent interest since the discharge physics is the same for a two-sided discharge between the waveguide walls and for a one-sided discharge on the target. The studies of discharges in the empty waveguide employed a multigrid electrostatic probe located in the center of the wide wall of the waveguide and a microwave probe with a magnetic pickup loop for recording the microwave field level in the waveguide.

Typical oscilloscope traces of the envelope of the microwave pulse from the microwave probe and of the collector current from the electrostatic probe are shown in Fig. 14.

It was noted that factors which distort the pattern for a pure SED (for example, plasma formation in the residual gas) appear more strongly with longer times after the discharge current begins. Consequently, in plotting electron current as a function of the field strength in the waveguide we have chosen the value of the electron current corresponding to a time 0.5 μsec after the electron current pulse begins on the oscilloscope traces (see Fig. 14).

From the dependence of the electron current I_p to the wide wall of the evacuated empty waveguide on the magnitude of the microwave electric field E_m (Fig. 15), it is clear that there is actually a threshold in E_m, $E_{m\,thr} \simeq 5$ kV/cm, beginning at which the interaction of the microwaves with the waveguide walls results in formation of an electron avalanche. Beyond the threshold, as E_m increases the value of I_p varies little over the interval 5-8 kV/cm and begins to fall when $E_m > 8$ kV/cm. The absolute value of the electron current density to the wide wall of the waveguide at the peak of the dependence on E_m reaches 10^{-4} A/cm^2.

In the experiments it was found that there is a definite delay τ_d between the introduction of microwave power into the waveguide circuit and the appearance of an electron current. This delay depends on the strength of the microwave electric field and is on the order of the duration of the microwave pulse at the threshold for existence of a discharge and reaches a minimum when $E_m \simeq 7$ kV/cm ($\tau_{min} \simeq 4\ \mu$sec).

The multigrid probe was used to analyze the electron energy. Figure 16 shows a typical variation in I_p when a retarding potential U_p is applied to the analyzer grid.

The very fact that a vacuum microwave electronic discharge develops with fields on the order of 5 kV/cm is certainly of interest. In fact, one of our main conclusions is that an electron avalanche can develop between two electrodes under conditions such that the electron motion is not cyclic and the resonant secondary-emission microwave discharge scheme is not applicable. These conditions are characteristic of the experiment described here. For $\omega \simeq 2 \cdot 10^{10}$ sec^{-1} and d ~ 3 cm, a microwave field on the order of $E_{m\,res} \simeq 10^5$ V/cm is required for resonant breakdown in the lower modes. A discharge actually begins at substantially lower fields (~5 kV/cm). At such low fields, only resonant modes of high order should develop ($k > 10$). However, the possibility of forming electron avalanches at such high modes is eliminated by the spread in initial velocities of the secondary electrons [for d ≃ 3 cm, $\omega \simeq 2 \cdot 10^{10}$ sec^{-1}, $E_m \simeq 5$ kV/cm, and $T_0 \sim 2$-4 eV condition (13) is clearly satisfied]. In addition, the measured discharge threshold is very close to that calculated with Eq. (32).

There is therefore reason to suppose that the discharge we have observed is the uniform polyphase secondary-emission discharge whose existence was predicted in Secs. 2 and 3.

An analysis of the measurement results, however, shows that a number of the phenomena observed in the experiments are hard to explain by means of our simplified model for formation of electron avalanches during the interaction of a microwave electric field with electrodes. In particular, it is unclear what causes the delay between the introduction of microwaves and breakdown. The shape of the electron energy spectrum derived from Fig. 16 is also of interest. A large fraction of the electrons detected by the probe have an energy which exceeds both the maximum oscillatory energy W and the maximum energy attainable in the microwave field, $W_{max} = 4W$. One could attempt to explain the "anomalous" tail in the distribution as a result of reflection of electrons from the chamber walls which was neglected in the simple model (i.e., some electrons accelerated to energies on the order of W_{max} are reflected from the waveguide wall and continue to gain energy in the microwave field as they move to the opposite wall). Special measurements of the coefficient η for a stainless steel surface under conditions close to our experiment were made at our request by Yu. A. Morozov's group at the M. I. Kalinin Leningrad Polytechnic Institute. According to their data the fraction of reflected electrons at primary electron energies close to W is 20%. The unexpectedly high value of η is an argument in favor of this mechanism for formation of a "tail" in the energy spectrum.

In order to explain the characteristic delay between the introduction of the microwave power into the waveguide and electron breakdown we suggest special measurements be made in the prebreakdown stage of the discharge.

Strictly speaking, in these experiments there is no unique evidence in favor of a secondary-emission origin for the electron current. Unfortunately, the design of the apparatus practically excludes the possibility of studying the interrelation between the secondary-emission properties of the electrodes and the threshold for a two-sided discharge. However, there is some indirect proof that the electron cloud in the volume of the waveguide is produced as a result of multiplication at the walls. Thus, a discharge does not develop when the inner surface of the waveguide is coated with a layer of soot. Heating the walls of the evacuated portion of the waveguide to 300°C for several hours leads to the same effect. The cutoff of the electron discharge in a "hot" waveguide is apparently due to a reduction in σ because of desorption of gases from the wall surfaces. The measurements made at the Leningrad Polytechnic Institute of the secondary emission coefficient of stainless steel plates confirm this assumption. Heating of stainless steel plates is in fact accompanied by a reduction in σ_b to values $\sigma_b \lesssim 2$ (at a temperature on the order of 300°C). According to the analysis of Sec. 3 a uniform polyphase secondary-emission discharge cannot exist when $\sigma_b < 2$.

When the sensitivity of the measurement apparatus was increased, it was possible to see that, almost immediately after microwave power was delivered to the waveguide, a "small" current appeared prior to the above-mentioned two-sided discharge current (Fig. 17). While the discharge current was abruptly stopped after heating or aging with microwave power, the "small" current was almost unchanged with heating or aging. The density of this current is $\sim 3 \cdot 10^{-7}$ A/cm^2. The "small" current was not cut off, unlike the "big" current, even at the largest power which could be obtained in this apparatus.

After the waveguide walls were coated with soot the "small" current decreased by several times but did not cut off completely. The "small" current appeared on the probe after about 1 μsec at a power of ~100 kW. As the power level was increased, the delay before its appearance decreased andreached 0.2-0.5 μsec at 1 MW. Evidently this current is a result of a two-sided discharge in the unheated portion of the device or of microscopic arcs at points of poor contact in the waveguide junctions.

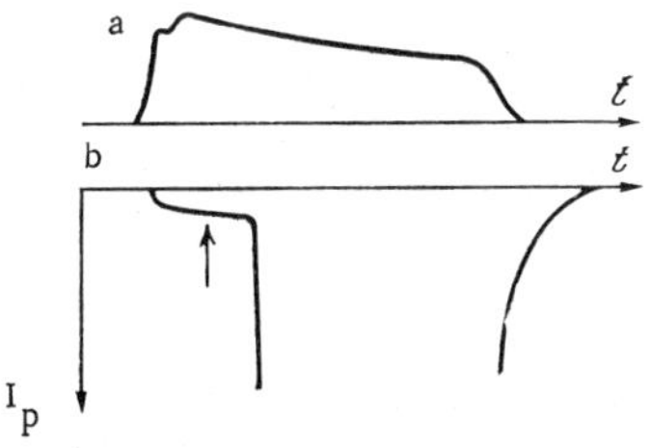

Fig. 17

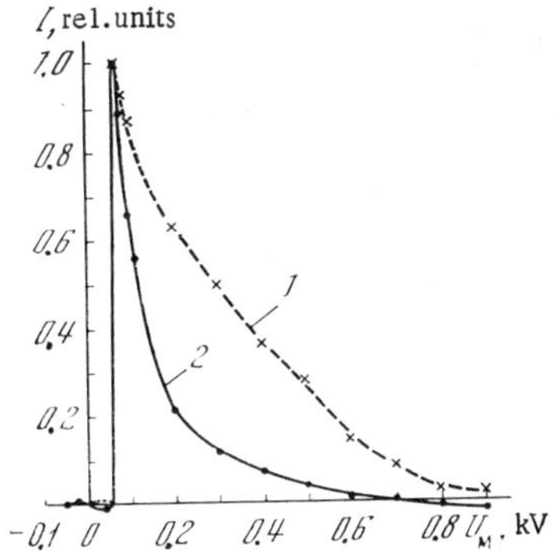

Fig. 18

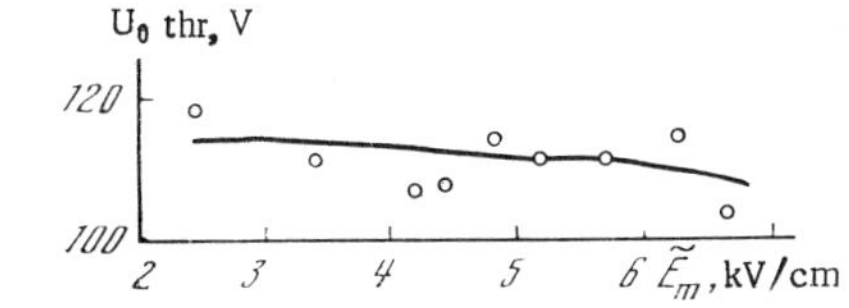

Fig. 19

Fig. 17. The "big" and "small" (arrow) currents: (a) the envelope of the microwave pulse; (b) probe current.

Fig. 18. The current to the probe (1) and the current coming directly from the target (2) as functions of the target potential.

Fig. 19. The threshold voltage as a function of the amplitude of the rf field.

Information on the background current is required for the correct interpretation of the later experiments.

11. Threshold Constant Restoring Field for a One-Sided Discharge (Applied Potential)

Heating the chamber walls or blackening them with soot creates conditions which inhibit formation of an electron avalanche between the walls of an empty waveguide. By placing metal plates in such a waveguide as shown in Fig. 13, it is possible to study the interaction of microwave fields with isolated surfaces (the case of a single surface, corresponding to Fig. 9b). The major advantage of such experiments compared with studies of discharges in an evacuated waveguide is the ability to change the plates. By using plates with different values of W_1 (the first critical potential was measured directly in the experimental device), it is possible to investigate the relationship between the breakdown threshold and the secondary-emission properties of the target.

The measurements showed that an electron avalanche is actually formed at a metal surface when microwaves interact with it. The occurrence of breakdown was detected by means of the multigrid probe and directly from the electron current to the plate. It seems that, as might be expected, two threshold values must be exceeded for a discharge to develop: one in the microwave electric field and the other in the restoring force on the secondary electrons. In the experiments a restoring force was created by applying a constant positive potential U_0 to the plate. The arrangements for applying a voltage to the target and measuring the discharge current are shown in Fig. 13.

The typical variations in the current to the multigrid probe and in the current from the target with the target potential [and, therefore, with the magnitude of the restoring force $f_r = eU_0/(d/2)$, where d is the distance between the wide walls of the waveguide] are shown in Fig. 18. These plots have been made for a microwave electric field level above the threshold value. From the figure it is clear that when the plate potential reaches a sufficiently high value, both the multigrid probe and the measurement circuit on the target record a jump in the electron current that is indicative of a surface discharge. Experimental plots of the value $U_{0\,thr}$ corresponding to this jump as a function of the first critical potential and of the microwave electric field strength have been made. As might be expected from Eq. (64), the measured values of $U_{0\,thr}$ are practically independent of the microwave field strength beginning with $E_m > E_{m\,thr}$ (Fig. 19) and increase almost linearly with the first critical potential (Fig. 20). The latter fact is evidence that the discharge has a secondary-emission character.

A single surface breakdown occurs with a restoring force that is much lower than that required for the first modes of a resonant discharge. Furthermore, at such low values of this force the criterion for "nonresonance" [Eq. (13)] is clearly satisfied. The electron energy distribution obtained from the electrostatic probe measurements has a "tail" at energies above W_{max} as in the case of an empty waveguide. The time delays between the introduction of the microwave power and the onset of the discharge depend on the magnitude

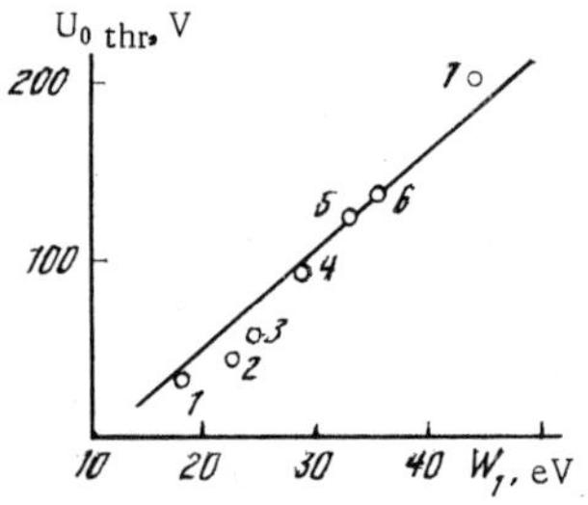

Fig. 20

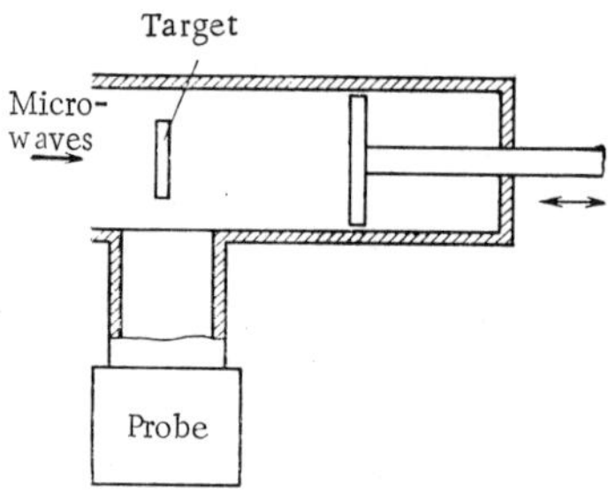

Fig. 21

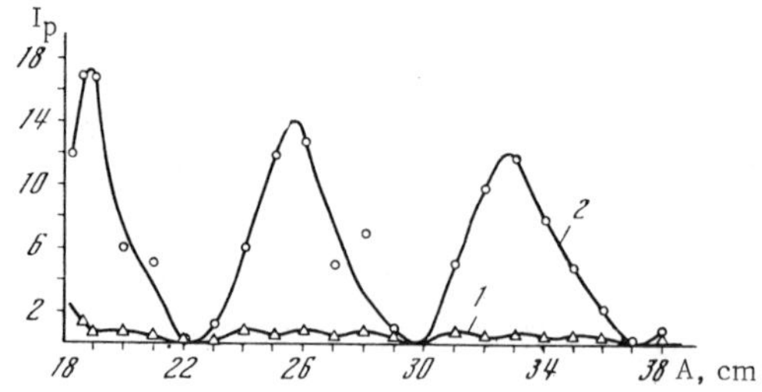

Fig. 22

Fig. 20. The dependence of the threshold potential on the first critical potential of the material: (1, 2) stainless steel; (3, 6, 7) duralumin; (4) copper; (5) platinum.

Fig. 21. A diagram of the experiment on discharges at a dielectric in a standing wave field.

Fig. 22. The current to the probe as a function of the distance between the target and a short-circuiting piston for two orientations of the LiF target relative to the microwave electric field: (1) $E \perp S$; (2) $E \parallel S$; P = 240 kW.

of the potential U_0 and are almost absent when U_0 exceeds $U_{0\,thr}$ by a few volts.

The results of these experiments on the effect of the target potential on the course of a secondary-emission discharge are important in two respects. First, it is interesting to know to what extent our theoretical ideas on the thresholds for development of a discharge are confirmed experimentally. Second, although we do not have a theory for a developing polyphase discharge, experimental material on the variation in the intensity of a discharge with the target potential is nevertheless of definite interest.

As for the first question, the experimental results have shown that we apparently have a sufficiently accurate idea of the physical processes which occur in the initial stage of the discharge. In fact, as we have conjectured, the discharge develops with a constant target voltage that is much lower than that required to produce the restoring force for a resonant discharge in the lowest modes. It is easy to calculate the constant field strength E_{const} required for an electron to be in flight for a specified time t_f:

$$E_{const} = \frac{2}{t_f} \sqrt{\frac{2mU}{e}},$$

where U is the potential corresponding to the energy of translational motion. Thus, for P = 1 MW, U = 330 V, and $t_f = T$, it is necessary to apply a constant electric field of 3.36 kV/cm. In the experiment, however, we observed onset of a discharge when $E_{const} \sim$ 30-120 V/cm, which is evidence for the existence of a uniform polyphase SED.

The experimental dependence in Fig. 20 was obtained by placing targets made of different materials in the waveguide: platinum, copper, several kinds of stainless steel, and aluminum, as well as plates coated with soot. The linear dependence of $U_{0\,thr}$ on W_1 (the first critical potential) is in agreement with the qualitative theoretical discussion of the uniform polyphase SED (see Sec. 6).

As for the answer to the question of how an increase in the target potential affects the intensity with which the surface is bombarded by electrons, some information can be obtained by studying the dependence of the current to a probe located opposite the target on the target potential (Fig. 18). From this graph it is clear that as the target potential is increased (if $U_0 > U_{0\,thr}$) the current to the probe decreases. This result could be regarded as a reduction in the intensity with which the surface is bombarded by electrons in a one-sided discharge as the restoring force is increased. However, there is not yet sufficient justification for this since the probe only detects electrons which have escaped the target. While electrons from the target may not reach the probe because of an excessively large restoring force, an intense discharge might nevertheless continue. The question of how the intensity of electron bombardment of the surface varies with the target potential can be correctly answered only by experiments to measure the current to the target itself.

12. The Interaction of a Microwave Field with a Dielectric in a Standing Wave Field for a Single-Sided Discharge

The preceding experiments with our equipment for observing an SED have dealt with the interaction of a microwave field with materials in a traveling wave field.

Experiments in the field of a standing wave are of interest for two reasons: on one hand, it is desirable to test the ideas on the physics of the discharge that have been obtained in traveling wave experiments under standing wave conditions; on the other hand, a standing wave has an important feature associated with the existense of a quasipotential.

Since a restoring force for the electrons must exist in order for a secondary-emission discharge to begin on an isolated object, it is natural to assume that the gradient force of a quasipotential may serve as such a force. It is thus appropriate to examine the role of a gradient force as a factor leading to the development of a secondary-emission discharge.

Clearly, if a standing wave field that increases with distance from the surface is established at the surface, then this field creates a restoring force that aids the production of an SED. If, however, the field decreases with distance from the surface, then this structure inhibits discharge formation since any electron that escapes the surface no longer returns to it to generate secondary electrons. It is true, on the other hand, that the escape of electrons causes positive electrostatic charging of the object, that is, the appearance of an electrostatic restoring force.

An experiment to elucidate the role of a gradient force was conceived in the following way. A metal piston was mounted in an evacuated rectangular waveguide (Fig. 21). By moving the piston without disturbing the vacuum it was possible to obtain any region of a standing wave at the target.

By observing the probe current for different strengths of the interacting electromagnetic field and placing the probe in various regions of the standing wave region, it is possible to evaluate the role of gradient forces. Thus, if the target is located at a node of the standing wave and the surface of the target is parallel to the electric field of the wave, then one surface of the plate will be acted on by gradient forces which remove electrons from the surface and inhibit discharge formation, while the opposite surface of the plate will experience conditions favorable to discharge formation since the gradient force is directed toward the surface.

If, however, the plate is located at an antinode of the microwaves, then gradient forces which remove electrons from the surface and inhibit SED formation will act on both sides.

Prior to the experiments, in order to verify that the surface of the plate was precisely subjected to the above-mentioned conditions and that the target itself did not distort the structure of the field near the target, we made cold measurements of the field structure at a low microwave power level. It follows from these measurements that a plane dielectric target does not so distort the distribution of the alternating electric field as to cause additional gradient forces to appear at the surface which would be significant compared with the forces caused by the standing wave field in an empty waveguide.

The apparatus is sketched in Fig. 21. The variation in the current to the probe from an LiF dielectric target was obtained for two orientations of the plate $S \perp E$ and $S \parallel E$ (Fig. 22). The power of the generator was 240 kW.

No effects due to Miller forces on the conditions for development of a discharge were observed in the experiment. In fact, if the main effect that returned electrons to the surface were the Miller force, then no discharge could occur at the antinode of the alternating field because at that point the Miller force is directed outward from both surfaces.

We are thus left to assume either that the discharge is an SED, but the restoring force is produced by an electrostatic field, or that a discharge on a dielectric is not an SED but a surface breakdown for which restoring forces are not important.

13. Threshold in the Microwave Field Level for a One-Sided Discharge

Discharges on Dielectric Targets

The study of SED's on dielectrics presents serious difficulties for the following reasons:

First, it is difficult to obtain information directly on the secondary emission characteristics of the material in this device. This information must be known in order to compare the experimental data with the theoretical predictions.

Second, as electrons leave the target, the target acquires an electrostatic charge which varies during the discharge. This charge changes the restoring force, which in turn changes the discharge conditions. Thus,

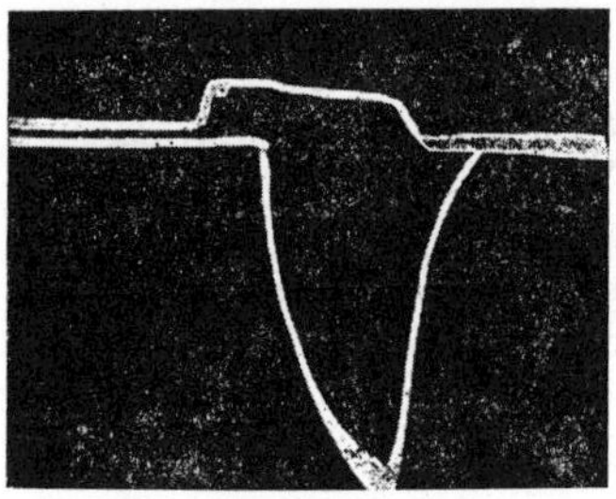

Fig. 23

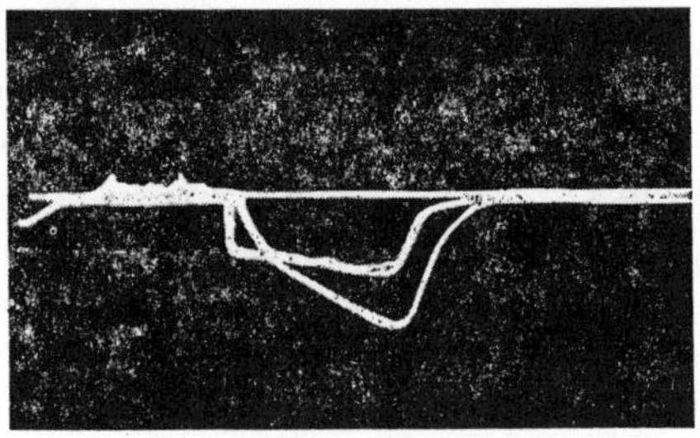

Fig. 24

Fig. 23. Oscilloscope traces of the current to the probe from the dielectric target (top) and of the envelope of the microwave power.

Fig. 24. Oscilloscope traces of the signals from the electric probe (larger amplitude) and from the photomultiplier.

in order to investigate SED's on dielectrics, it is necessary to have a method for continuously monitoring the potential on the surface of the dielectric, as well as a method for determining σ.

Third, it is difficult to correctly measure the current arriving at the surface of the dielectric. The current which arrives at measurement probes located away from the target only provides an indirect characterization of the processes which occur on the surface of the target itself.

Fourth, it is impossible to arrange the available apparatus so that the microwave field does not envelop the edges of the target. As shown in Sec. 5, this is a fundamental condition for any determination of the discharge threshold.

These difficulties are comparatively easy to overcome with metal targets. Since the physics of the SED effect is the same on metals and dielectrics, our studies of the SED were mainly done on metal targets. The experiments with dielectric targets were only of subsidiary importance.

The following equipment was used in the experiments on one-sided discharges on dielectrics in the field of a traveling wave: a multigrid probe, a microwave probe with a magnetic pickup loop, and a photomultiplier located in the narrow wall of the waveguide opposite the end of the target to detect the luminosity of the target.

The current on the oscilloscope traces 1 μsec after the beginning of a discharge was used in the graph as the current from the target.

Some typical oscilloscope traces of the envelope of the microwave pulse from the microwave probe and of the collector current to the probe are shown in Fig. 23. Traces of the electric probe and photomultiplier signals are shown in Fig. 24. The target material is glass. We may conclude from these traces (Fig. 24) that during the time that the SED process is comparatively constant (if we consider the photomultiplier signal), the current to the probe during the microwave pulse changes substantially. This is evidence that the ability of the current to reach the probe is improved because of ionization of both the residual gas and the gas released by electron bombardment of the target. In a plot of the probe current as a function of the microwave power level in the waveguide for glass (Fig. 25a), we can isolate two characteristic parts of the curve with different slopes. Evidently, these two regions involve different physics. We shall discuss this in more detail after describing the experiments with metal targets. The dependence of the probe current on the angle of rotation of the target plane relative to the E-field of the wave was measured for glass and LiF targets (Fig. 25b and c).

Current maxima near angles of 0° and 180° are characteristic of both dielectrics. This might be expected from the theoretical discussion of Sec. 5 because at small angles (β = 15°-20°) of inclination of E to the target plane the smallest value of $W_1(\beta)$ and, therefore, the largest of $\sigma(\beta)$, is achieved. This effect may be regarded as confirmation of the secondary-emission nature of the discharge.

Discharges on Metallic Targets

It is much more convenient to study SED's on metallic targets than on dielectric targets because it is possible to monitor the magnitude of the restoring force and it is relatively easy to measure the secondary-emission properties of the surfaces directly in the apparatus.

The experiment was done as follows. The target was positioned perpendicular to the lines of force of the microwave electric field ($E \perp S$). A threshold potential was applied to the target and the current to the opposite

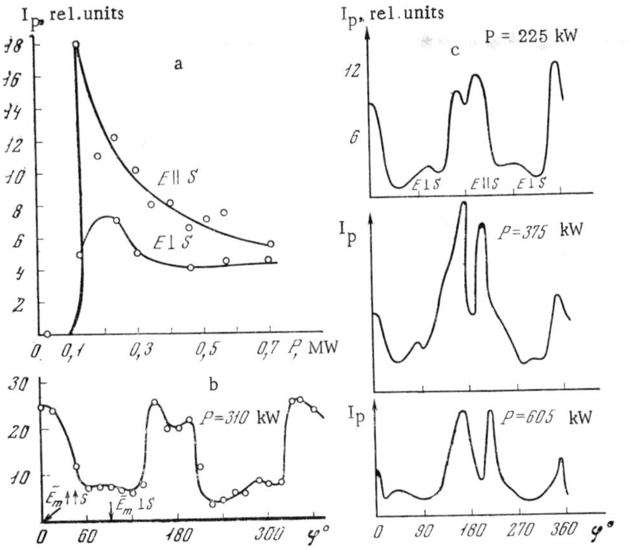

Fig. 25. The current to the probe as a function of the microwave power level (a) and of the angle of rotation (b, c): (a), (b) glass target; (c) LiF target (temperature of the waveguide walls 300°C).

electric probe was measured; the electron current leaving or arriving at the target was also measured directly galvanically. At the same time the microwave probe was used to record the microwave power level in the waveguide. In the wall opposite that containing the electric probe, there was an electron gun. It was used to measure the first critical potential W_1/e of the target without removing the target from the waveguide.

A diaphragm was placed in the probe port to limit the angle of view of the probe so it could only "see" particles coming from the center of the target. In order to reduce the effect of the walls, during the entire experiment the waveguide walls were kept at ~300°C so that a discharge could not develop between the walls.

A secondary emission discharge can be identified in terms of the change in the electron current to the wide wall of the waveguide and in terms of the current to the test sample. The signal from the probe is shown in Fig. 26a. As the target potential is gradually increased, the probe current initially appears with a large delay relative to the beginning of the microwave pulse. As the voltage is increased the delay is reduced. We shall consider the threshold potential to be that at which the delay is 4 μsec.

The increase in the probe signal during the microwave pulse is somewhat surprising. It might be supposed that the increase in the signal is a result of gas released by bombardment of the surface. Part of this gas is ionized. This leads to neutralization of the space charge and thus to a rise in the current. In fact, after several hours of aging the surface with a discharge current, the surface was outgassed and the probe signal took the form shown in Fig. 26b. The sharp peak at the beginning of the probe signal is, of course, explained by the fact that at the moment of breakdown the space charge has not grown sufficiently to limit and defocus the electron flux moving from the target to the probe. Figure 27a shows the dependence of the probe current* on the microwave power for a stainless steel target. The initial current arriving at the probe 0.5 μsec after the beginning of the microwave pulse was plotted in these graphs. As a rule, if the current at 2-3 μsec after the beginning of the microwave pulse was plotted, then the power dependence will be the same as for the initial current. Similar dependences were observed for targets made of other metals.

When comparing the experimental results with the calculations it is necessary to note that the calculations were done for unbounded electrodes with a uniform field between them. During the gradual rise in the microwave field the discharge must simultaneously break out over the entire surface at the moment the threshold field is attained such that $\langle \sigma \rangle = 1$.

*This characteristic shape appears only with sufficient discharge aging. The dashed curve was obtained after brief aging. Similar effects were observed with other targets as well. Evidently, the gas that is released and the plasma formed from it cause the apparent rise in the target current as the power is increased.

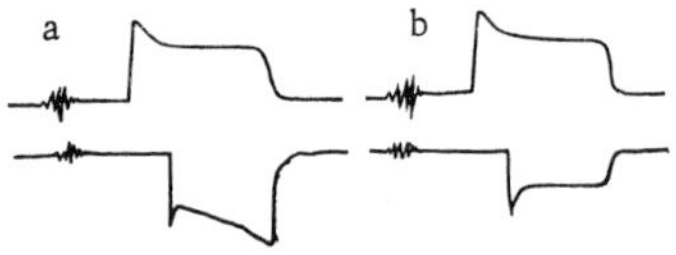

Fig. 26

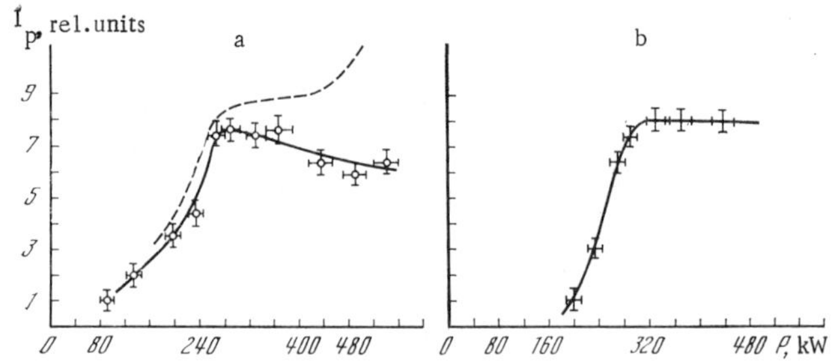

Fig. 27

Fig. 26. The probe signal for targets without (a) and with (b) aging. The upper trace is the envelope of the microwave pulse.

Fig. 27. The probe current as a function of the microwave power level: (a) clean target; (b) edge of the target coated with soot. The dashed curve is the current from an unaged target.

The experiment was actually carried out on a finite target. Evidently, the electric field of the wave is greater on the edges of the target than at the center (field enhancement at a sharp point). Thus, when the power level is gradually increased a discharge should initially appear at the edge of the target. As the microwave power is increased further the region in which a self-sustaining discharge can exist will expand until it reaches the center of the target. It might be possible to establish the threshold microwave power level by recording the appearance of a discharge at the edges of the target; however, it is difficult to measure the field at this location. Consequently, the experiment was set up so as to record the time when $\langle\sigma\rangle = 1$ was reached at a part of the target where the field is known exactly. One such part is the center of the target where the field is exactly equal to the field in the empty waveguide. The viewing angle of the probe was chosen so that the probe could "see" only the center of the target. A determination that conditions were such that a self-sustaining discharge had appeared at the center of the target was based on the following facts.

If the microwave field at the center of the target is below threshold $E_{cen} < E_{thr}$, but somewhere on the target where the field was higher a discharge appeared, then the current density falling on the center of the target, I_{cen}, has the dependence

$$I_{cen} = \frac{I_0}{1 - \langle\sigma(E_{cen})\rangle}, \tag{70}$$

where I_0 is the current carried to the center from a discharge at the periphery by the "thermal" velocities along the surface of the target, by space-charge repulsion, and so on, as well as by the background current in the waveguide.

As the power level P is increased the current I_0 will increase because the self-sustained discharge region approaches the center. In addition, $\langle\sigma(E_{cen})\rangle$ will increase with E_{cen}. Consequently, as the power rises the current at the center of the target will increase although E_{cen} remains below E_{thr}. But as soon as E_{cen} exceeds E_{thr}, the dependence of I_{cen} on P changes, since in this case I_{cen} will be determined by another factor, specifically, the stabilizing effect of the space charge.

Therefore, the power level at which $\langle\sigma\rangle = 1$ at the target center can be found from the experimental dependence of I_{cen} on P. The threshold power which creates the threshold field at the target center is that power at which the $I_{cen}(P)$ curve has a deflection. As was to be expected, the experimental $I_{cen}(P)$ curve (see Fig. 27) has two characteristic sections with a noticeable deflection separating their slopes.

In order to further ascertain the validity of these considerations, the edges of the target, where the field is most different from that at the center, were coated with soot on which a discharge can certainly not occur. At different points on the clean surface the microwave field became more uniform and the discharge at different points developed at values of E that were closer to one another. As a result the curvature of the left side of the curve actually increased sharply (see Fig. 27b). This confirmed the validity of our picture of how the discharge is ignited.

In order to determine the threshold field for a given material it is therefore necessary to measure the power at which the $I_{cen}(P)$ curve has a deflection. The field strength in the empty waveguide corresponding to

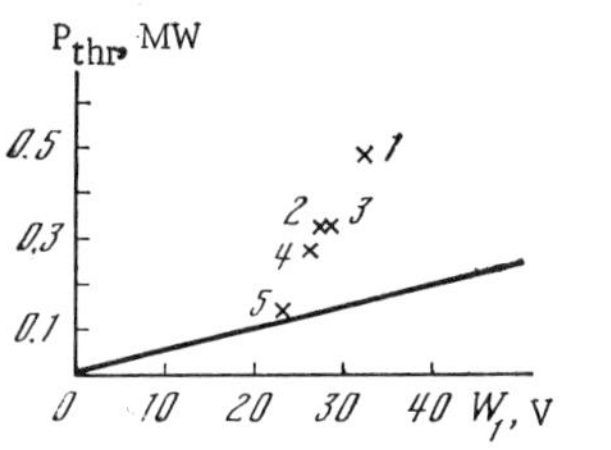

Fig. 28

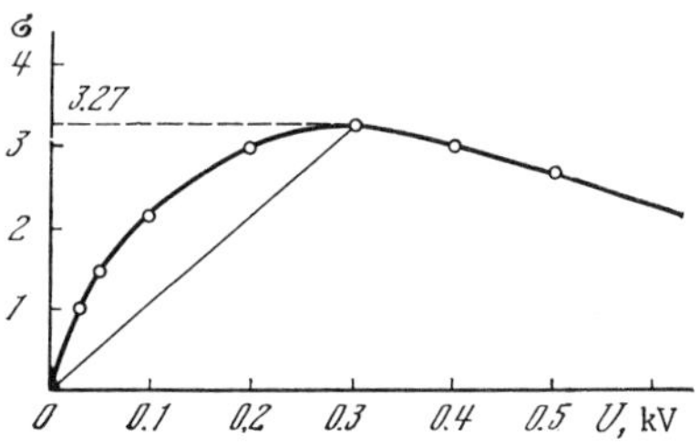

Fig. 29

Fig. 28. The magnitude of the threshold power as a function of the first critical potential: (1) platinum; (2) copper; (3)-(5) stainless steel I-III, respectively. The straight line corresponds to $W_k = 0.92 W_1$.

Fig. 29. The secondary-emission characteristic of a copper target.

this power is also the threshold field strength for the target material.

If we plot the threshold powers found in this manner for various materials as a function of the first critical potential W_1 (Fig. 28), then we find that our result is in full qualitative agreement with the theory. The threshold power rises along with the first critical potential. The same figure shows the theoretical dependence of the threshold power on the first critical potential for substances where σ is proportional to the energy of the primary electrons up to $\sigma = 3.76$.

Comparing the experimental points with the theoretical dependence, we see that, as should be expected, the experimental points lie above the theoretical curve. This is so because, for the metals we have studied, the linear portion of the energy dependence of σ does not extend to 3.76, where the theory is applicable, but only to smaller values, and it becomes a horizontal line at lower energies.

This means that with such materials a discharge will begin at higher power levels, as we can see in Fig. 28. Naturally, it would be interesting to compare this with the results of a more precise theory in which the actual energy dependence of σ is replaced by a curve consisting of two broken lines. However, in order to do this it is necessary to plot the entire energy dependence of σ (and not just determine the value of W_1) directly in the waveguide for all the materials that are tested.

It turns out that this is a very difficult problem. The main obstacle is the fact that the secondary-emission curve (see Fig. 4) changes during aging with microwave radiation (this may be ascertained from the recorded value of W_1). Nevertheless, it was possible to identify two materials (copper and stainless steel) whose secondary emission characteristics (judging from the weak variation in W_1) remained stable after preliminary aging for the rather long time necessary to obtain a complete secondary-emission characteristic with the microwave field turned off.

From the secondary-emission characteristic of copper (without any special processing) obtained on this apparatus (Fig. 29) it follows that $\sigma_b = 3.27$, $W_b = 300$ V, $\gamma = 1/5$. Using Fig. 6, taking the curve that is closest in γ ($\gamma = 1/10$), we find $\beta_{thr} = 3.25$ for copper, which corresponds to $W_{thr} = W_b/\beta_{thr} = 92$ eV. The threshold power needed to produce a discharge is found by employing the relationship between the power propagating through the waveguide and the oscillatory energy of the electrons [for the wavelength and waveguide used $2.5 \cdot 10^{-4} P\,(\text{watts}) = W\,(\text{eV})$]. Thus, the calculated threshold power $P_{thr} = 370$ kW. The experimental value is $P_{thr} = 320$ kW.

A summary and comparison of the experimental thresholds and the thresholds calculated using the method developed in Sec. 3 is given in Table 1.

We note that the experimentally determined values of the power for the first threshold are somewhat lower than the theoretical. This may be explained by the fact that at low electron energies the target actually has a larger value of σ (see Fig. 29) than that used in the calculation (the straight line).

With this generator it was not possible to obtain a sufficiently high microwave field in a traveling wave to cut off a uniform polyphase SED by reaching the second threshold (on any of the targets). An exception was a discharge between the walls of the waveguide where σ_b was reduced by heating the walls to 300°C. In order to

TABLE 1

Material	Threshold number	P_{thr}, kW	
		experiment	theory
Copper	1	320	370
Stainless steel I	1	275	330
Stainless steel II	1	320	330
Copper	2	1400	2400

increase the microwave field an experiment was done in a standing wave. Then a discharge on a copper target was cut off as predicted by the theory. The field strength corresponded to a traveling wave with a power of about 1.4 MW.

A significant difference was observed between the measured value of the second threshold microwave field and the calculated value. Possible reasons for the reduction in the power required to reach the second threshold include the Miller force and space-charge effects.

In conclusion, we note that a more complete comparison of the computed thresholds with those determined experimentally on a device of this type could be made using an electronic circuit that allowed the complete secondary-emission characteristic to be taken over a shorter period of time comparable to the microwave pulse duration.

Another way to do this is to substantially improve the vacuum conditions in the equipment and choose target materials whose secondary-emission characteristics will remain constant after microwave aging for the time required to measure the characteristics.

14. Relation between the Threshold with Respect to the Microwave Field and That with Respect to the Restoring Field

In the experiments of Sec. 11 the threshold power was defined as the power at which the dependence of I_p on P changes. A fairly good agreement between the experimental and theoretical data was obtained in that case.

After we had determined the threshold potentials for a number of materials from the inflection point (see Fig. 28) and found the potentials $U_0 thr_1$ (see Fig. 20) at which the current from the target begins to rise rapidly for the same materials, it became possible to compare how the potentials $U_0 thr_1$ and the powers corresponding to the inflection points for various materials are related to one another.

The $U_0 thr_1$ and P_{inf} for each material that was tested are plotted in Fig. 30. The relationship between $U_0 thr_1$ and P_{inf} that follows from the following analysis is plotted on the same graph.

Let microwaves at a power that is clearly in excess of the threshold act on a substance. In this case the discharge can maintain itself if $U_0 < eE^2/2m\omega^2 = 2.5 \cdot 10^{-4}$ P since, although part of the electrons with energies greater than eU_0 leave the target, the remaining electrons with smaller energies can keep $\langle\sigma\rangle = 1$. The situation is different if a threshold power acts on the target. In order to keep $\langle\sigma\rangle = 1$ in that case, all the emitted electrons must return to the surface. This, however, is possible only if $U_0 \geq 2.5 \cdot 10^{-4} P_{inf}$. But $U_0 thr_1$ is the minimum value of the constant potential U_0 at which a discharge is possible. Consequently for the threshold microwave power the condition

$$2.5 \cdot 10^{-4} P_{inf} = U_0 thr_1$$

must be satisfied. This corresponds to the straight line in Fig. 30. From the figure it is clear that the inflection points lie close to the theoretical curve. This confirms our assumption (see Sec. 13, Paragraph 2) that the inflection point is the threshold power.

We have therefore found the relationship between the threshold with respect to the microwave field (see Sec. 13) and the threshold with respect to the restoring field (see Sec. 11). Thus, the results of the two sections confirm one another.

15. "Self-Charging" of a Target in a Microwave Field

Since a constant electric field plays a large role in the discharge process (it has an effect on the intensity of the discharge and also creates the conditions for a discharge), it is important to examine how the potential of an isolated body changes if it is acted on by a high-power microwave field.

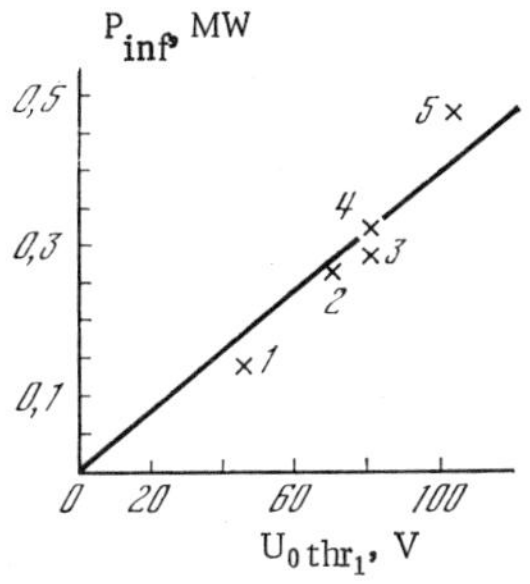

Fig. 30

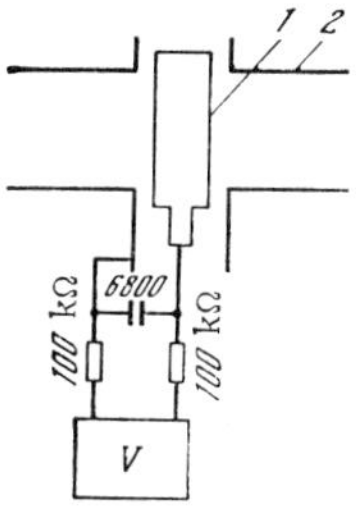

Fig. 31

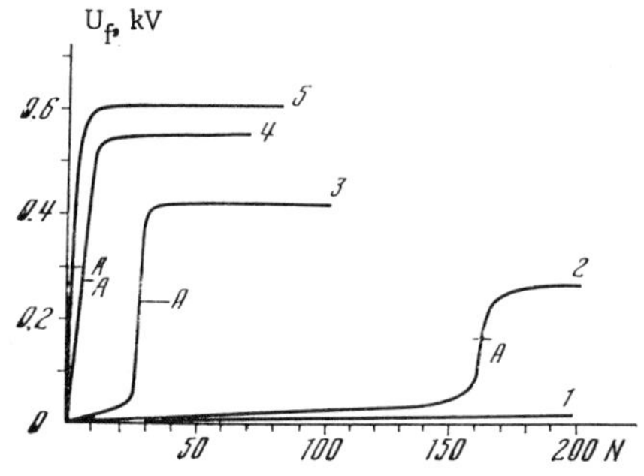

Fig. 32

Fig. 30. The relationship between the power P_{inf} corresponding to the "inflection point" and the threshold potential. ×, experiment; straight line, theory.

Fig. 31. A diagram of the apparatus used for measuring the floating potential of the target: (1) target; (2) waveguide; (V) electrostatic voltmeter.

Fig. 32. The time variation (t = 8N μsec) of the potential of a stainless steel target for various microwave power levels: $P_5 > P_4 > P_3 > P_2 > P_1$; A denotes the half height points.

A study of the dependence of the current from the target on the potential applied to it when it is subjected to microwaves (see Fig. 18) indicates that, if the target is charged to a potential equal to $U_0 thr_1$ and the charging source is removed, then the target is charged by the action of the microwave field to the potential $U_0 thr_2$ at which curve 1 intersects the abscissa a second time.

In fact, when $U_0 > U_0 thr_1$, $\langle\sigma\rangle > 1$ and part of the electrons can leave the target. After this the target is positively charged and it is difficult for more electrons to escape since the force which impedes their escape has increased.

The larger the positive potential is, the smaller the number of electrons which can overcome the electrostatic barrier and, on the other hand, the greater the number of electrons which will be attracted to the target from outside (in free space there are always some background electrons). At some potential there will be a balance between arriving and escaping electrons. The charging process ceases when the potential which we have denoted by U_{thr_2} is attained.

Some experiments on "self-charging" were performed and yielded just the result to be expected on the basis of the qualitative discussion given in Sec. 7 (see Fig. 12).

The circuit used for measuring the floating potential of the target is shown in Fig. 31. Unlike the preceding experiments, this measurement circuit does not include an oscilloscope as an indicator. This was done in order to reduce the effect of the measurement operation on the process as much as possible since the electrostatic voltmeter used in this circuit does not require a current, as does an oscilloscope.

The capacitance C = 6800 pF in parallel with the target served to reduce the effect of systematic changes in the capacitance of all the circuit elements during the experiment and, at the same time, made it possible to greatly extend the time for electrostatically charging the target. It was found that on charging to a certain potential during a microwave pulse, the condenser would not noticeably discharge during the pause between microwave pulses. This meant that a number of successive microwave pulses separated by periods of "silence" produce the same charging as a single, long, continuous microwave pulse of the same total duration.

The electromagnetic generator could emit pulses at a low repetition rate. This made it possible to follow visually on the electrostatic voltmeter the voltage by which the target potential changes during each 8-μsec pulse.

It is evident that were the target completely isolated from any measurement devices and from the auxiliary condenser, the potential change would take place with the same dependence on the duration of the microwave power as that observed in the experiment but over a time 6800 pF/C_M times smaller (where C_M is the capacitance of the target).

The experiment was conducted as follows. A sample of a given material was wiped with alcohol and then placed in the waveguide perpendicular to the lines of force of the microwave electric field. The temperature

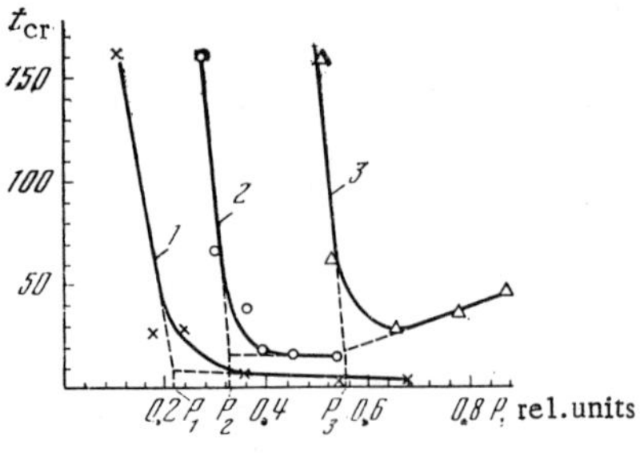

Fig. 33

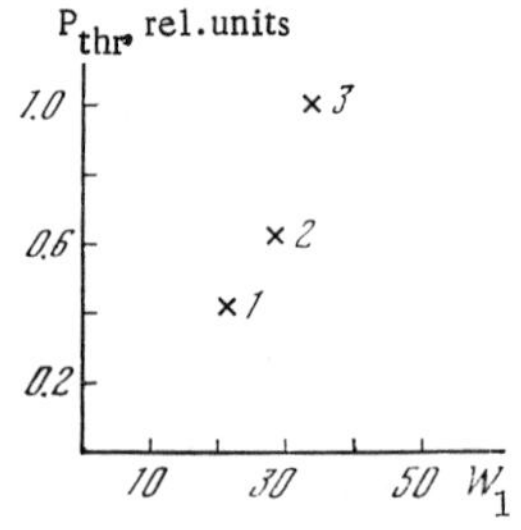

Fig. 34

Fig. 33. The microwave interaction time required for onset of a self-sustaining electron discharge at various microwave power levels for different materials: (1) stainless steel; (2) copper; (3) platinum.

Fig. 34. The threshold power for appearance of a discharge as a function of the first critical potential of the material; same notation as in Fig. 33.

of the waveguide was kept at about 300°C. After a vacuum of roughly 10^{-6} torr was reached in the apparatus, the target and waveguide were aged with a microwave power of 0.1-1 MW. Then the target potential was brought to zero and the first microwave pulse at a given power level was applied. After each pulse the target potential was noted and the dependence $U_f = f(N)$ (N is the number of the pulse) was constructed.

Regardless of the metal that was placed in the waveguide, the general character of the dependence of the target potential U_f on the number of pulses was the same.

The variation in the floating potential U_f of the target with the duration of the microwave interaction is shown in Fig. 32. The experiment was done with a stainless steel target. Each curve was taken at a fixed power. The graphs show that as the microwave power acts on the target its floating potential at first increases gradually. Whereas in the first few "shots" the potential increases linearly with the duration of the microwave interaction, later on the process speeds up and acquires an avalanchelike character. At a certain potential the charging processes ceases and the target remains at this potential throughout the rest of the microwave interaction. From the family of curves in Fig. 32, it is evident that the higher the level of microwave power is, the more rapidly the process becomes avalanchelike. However, this is true only up to a certain power beyond which it is impossible to achieve more rapid charging of the target. Sometimes the rate of charging even decreases slightly when the power is increased further.

It is interesting that different powers require different interaction times for the avalanche process to occur.

Graphs of the critical time required for the avalanche process to develop, t_{cr}, as functions of the power for three materials (stainless steel, copper, and platinum), which differ from one another in their first critical potential, are plotted in Fig. 33 (t_{cr} corresponds to the point A of Fig. 32). The curves for the different materials are similar to one another and have a characteristic bend. It is significant that the larger the first critical potential of the material is, the higher the power at which the bend occurs. Thus, stainless steel has a first critical potential of $W_1/e = 21$ V, copper has 28 V, and platinum has 32.5 V. Figure 34 shows the dependence of the threshold power on the secondary-emission characteristic of the material W_1.

The maximum potentials to which targets made of these materials can be charged by microwaves are plotted in Fig. 35 as functions of the incident power. From these graphs it is evident that the maximum target potential depends only weakly on the power, and once it has reached a certain value it hardly increases with the microwave power.

We now discuss these results.

Almost all the data in this section are basically in good agreement both with the qualitative ideas on the process and with the experimental results given in Sec. 13, Paragraph 2. It is easy to explain the dependences given in Fig. 32. In the beginning, at small t, the target is bombarded by background electrons. Since $\langle\sigma\rangle > 1$ the number of secondary electrons that escapes is greater than the number that arrives, so the target acquires

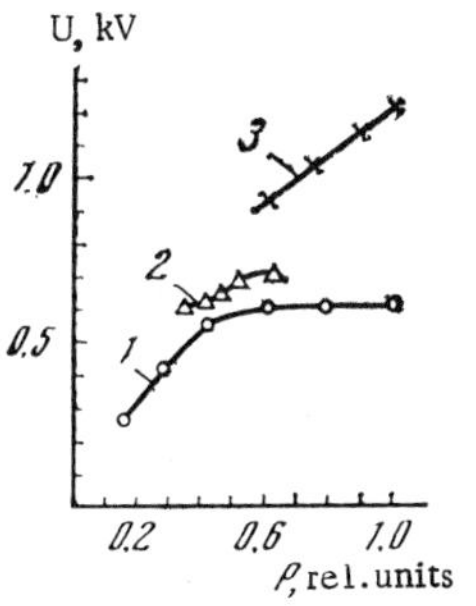

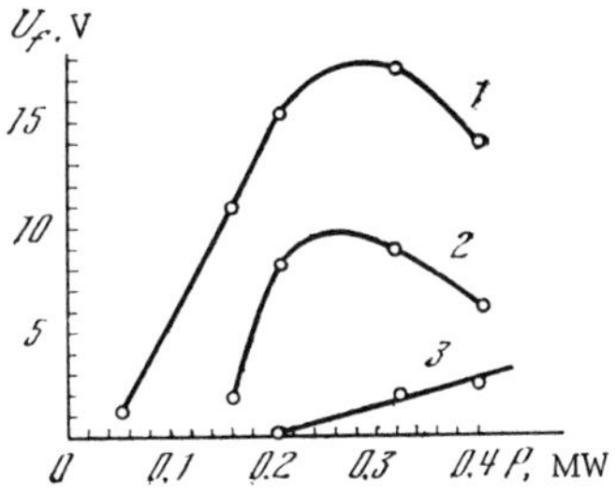

Fig. 35 Fig. 36

Fig. 35. The maximum potential to which a target is charged as a function of the microwave power; notation as in Fig. 33.

Fig. 36. The floating potential of a duralumin target as a function of the microwave power ($U_f < U_{0thr_I} > 50$ V). The curves characterize the reduction in the floating potential as the surface is cleaned: (1) 2 h aging at a pulse rate of 0.3 Hz; (2) 4 h at 0.3 Hz; (3) 4 h at 1 Hz.

a positive floating potential. With each microwave pulse the target becomes more charged until its potential reaches U_{0thr_I}. At that time the conditions arise for ignition of a self-sustaining uniform polyphase SED. The discharge starts and the number of electrons bombarding the surface increases sharply while the escape of electrons from the surface increases accordingly. This leads to a sharp jump in the target potential (see Fig. 32).

The increased restoring force causes a reduction in the number of escaping electrons. The rate of charging is reduced. When the potential becomes so high that the numbers of escaping and arriving electrons become equal, the target potential becomes stable. The rather high steady-state potential of the target is noteworthy (see Fig. 35).

The potential continues to increase while the electrons that are leaving the surface of the target can overcome the potential barrier created by the electric field of the target. The fact that the platinum target was charged up to a potential of 1.2 kV indicates that electrons with energies of 1.2 keV were emitted on its surface. This is considerably greater than the energy that an electron can receive in the field at this microwave power. The maximum energy with which an electron can leave the wall at this power is 0.35 keV.

In order to explain the existence of electrons with energies on the order of 1.2 keV, it must be assumed that these electrons have undergone at least one elastic collision with the target.

The fact that the platinum target was charged to 1.2 kV and not higher does not contradict the statement that still more energetic particles can exist in the discharge, having undergone not one, but several, elastic collisions. It is only necessary to note that a potential of 1.2 kV is established because it is just at this potential that a balance is established between the number of arriving and departing particles and not because there are no more energetic electrons. It can be stated that for a discharge on a surface in free space the potential will be still higher. It is significant that the charging of the target to a high potential confirms the presence of fast electrons in the discharge (fast electrons were observed during an investigation of the energy spectrum of the electrons that participate in a discharge in a waveguide).

The fact that the floating potential of the target increases slowly (or not at all) after some value of the microwave power acting on a given target is reached (see Fig. 35) confirms that the fraction of fast electrons in the discharge is reduced as the power increases. This circumstance was noted during a study of discharges in an empty waveguide. In addition, a reduction in the mean relative energy of the electrons at increased powers was predicted theoretically (see Fig. 7).

It might also be assumed that the slow rise in the potential with the microwave power is associated with the fact that in general the current to the target does not rise with the power. And since the fast electrons are a part (elastically reflected) of the total current and proportional to it, the number of fast electrons will also rise slowly with the power.

A comparison of this explanation with our results on the dependence of the discharge current on the power level at a target with a fixed potential (see Fig. 27) and in an empty waveguide (see Fig. 15) allows us to draw the important conclusion that an increase in the microwave electric field above threshold does not lead to a proportional increase in the current density to the target but, rather, that the current density increases slowly or may even be independent of the power.

The summarizing data of Fig. 34, which are the result of an analysis of the plots of Fig. 33, show how the threshold power for "self-charging" of the target varies. The deflection points in the curves of Fig. 33 are taken as the threshold powers. As in Fig. 28 (which summarizes the experiments on targets with an externally applied potential), the larger the first critical potential is, the larger the power at which a discharge develops on the target.

It should be noted that "self-charging" of the target from 0 to $U_{0\,thr_2}$ only occurs when the current curve (see the dashed curve of Fig. 18) does not cross the abscissa as U_0 changes from 0 to $U_{0\,thr_1}$ (that is, when the current is positive in this range). This can be observed only when the target has not been aged for long. After protracted aging the current curve, as shown in Fig. 18, goes negative for $0 < U_0 < U_{0\,thr_1}$ and the target charges itself only to $U_f \sim 3$-30 V.

However, even in this case "self-charging" of the target from $U_{0\,thr_1}$ to $U_{0\,thr_2}$ can be observed if the target is initially charged by an external source to a potential of $U_{0\,thr_1}$. After this the potential begins to rise randomly up to $U_{0\,thr_2}$ in a manner similar to that described above.

In Sec. 7 it is shown that a self-sustaining discharge begins on an isolated target only if the floating potential reaches $U_{0\,thr_1}$. This could not be reached on all the materials that were studied. This became more difficult as the target was aged in a microwave field because, on one hand, W_1 was raised [and, therefore, $U_{0\,thr_2}$ as well; see Eq. (64)] and, on the other, as W_1 was raised the floating potential U_f for a given microwave field level was reduced [see Eq. (66) together with the graph of Fig. 11].

A comparison of the experimental values of U_f after 2 h of aging and after 4 h (curves 1 and 2 of Fig. 36) confirms the expected reduction in the floating potential of a target with a "cleaner" surface.

One curious fact was the extremely rapid, in a fraction of a second, "contamination" of the surface, even when the residual gas pressure was 10^{-6} torr. This can be seen by comparing the values of U_f (curves 2 and 3 of Fig. 36) reached by the target for microwave pulse rates of 0.3 and 1 Hz.

One of the major questions in studies of SED's is how the current density to the target depends on the energy density of the microwaves. In the experiments with a target that had an externally applied potential (see Sec. 11), it was found that after a certain threshold power the electron current leaving the target ceases to grow with the microwave energy density. Since the magnitude of the escaping current was compared at various powers but at the same voltage (i.e., at the same restoring force), the saturation of the escaping current was evidence of an effective saturation of the electron current bombarding the target. The same result was obtained in the study of discharges in an empty waveguide (unlike the results of the target experiments, this was a direct result). If we consider the rate of charging of targets at different microwave power levels (see Fig. 33), then we verify that in this experiment, after a certain threshold power, the rate of charging stabilizes and remains at that level despite the higher power level. This may indicate that the rise in the current to the target has ended. However, a final answer to the problem of how the electron current bombarding the target varies with the microwave field and with the magnitude of the restoring force can only be given by an experiment in which the current to the target itself is measured.

CHAPTER III

FULLY DEVELOPED POLYPHASE DISCHARGES

16. Direct Measurement of the Current Density to the Target

In the previous chapter we obtained information on the initial stage of a uniform polyphase SED and determined the thresholds with respect to the restoring force and with respect to the magnitude of the microwave field. Along the way some data were obtained on the developed discharge:

(a) as the restoring force is increased the current density of the discharge decreases (see Fig. 18); and

(b) as the microwave field is raised above threshold the discharge current becomes saturated (see Figs. 15, 27b, and 25).

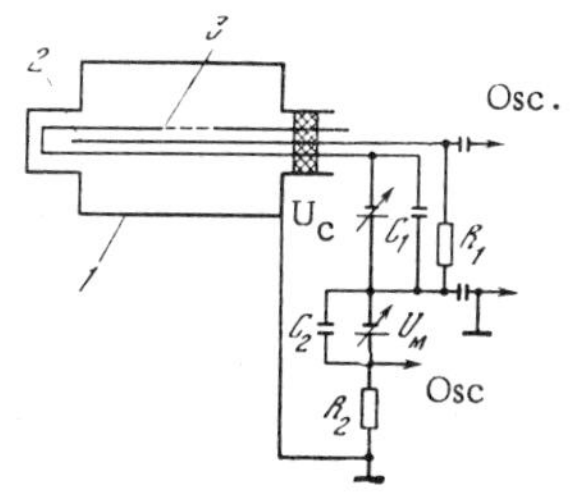

Fig. 37

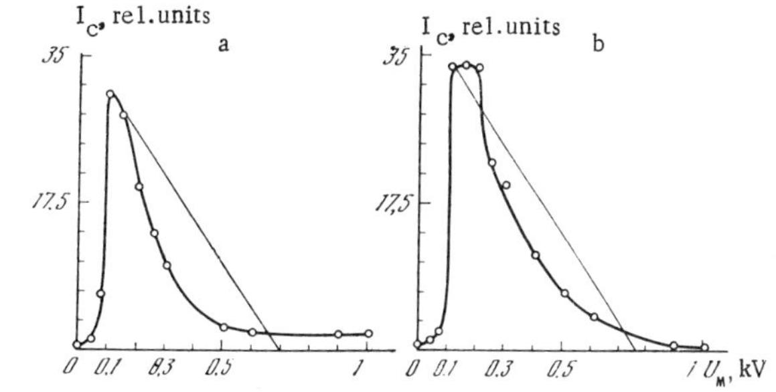

Fig. 38

Fig. 37. The target probe placed in the waveguide and the electrical circuit for the probe supply: (1) waveguide; (2) collector; (3) holes.

Fig. 38. The collector current as a function of the potential of the target probe (U_C = 200 V): (a) P = 363 kW; (b) P = 540 kW.

However, the discussion of the nature of the dependence of the discharge current on the microwave field level and on the restoring force was based on measurements of the current coming from the target. This is not completely correct since the current to the target and the current from the target may depend differently on the same factors. In order to measure directly the current density to the target we built a target probe. Figure 37 shows a longitudinal cross section of this probe and the arrangements for the electrical measurements.

The target was made of stainless steel and was the same size as that used previously. Inside the target there was a cavity within which was mounted a collector plate of thickness 0.2 mm that was galvanically isolated from the case and coated with soot to reduce secondary-emission effects. In the wall of the target case, at the location in the center of the waveguide where the electric field of the microwaves is greatest, 25 staggered holes were drilled opposite the collector over a total area of 1 cm^2. The diameter of the holes was 0.1 mm and the thickness of the target case wall at this point was 0.15 mm. The holes were made so small in order to minimize the effect on the discharge. A hole distorts the field near the surface over a distance on the order of the hole diameter Φ. Thus, in order that the effect of the distorted field on the energy of an electron be negligible it is necessary to have $\Phi \ll L$, where L is the distance moved by an electron in a single period of the field. At a microwave power of 1 MW, L ~ 2 mm.

The current incident on the collector as a function of the target potential was obtained for two values of the microwave power (Fig. 38). The collector potential is 200 V. From these curves it is clear that as the target potential U_0 is increased the electron current bombarding the target surface falls by an order of magnitude by the time $U_0 \simeq 500$ V. Therefore, the conclusions derived from observations of the current leaving the target are confirmed: increasing U_0 beyond $U_{0\,thr_1}$ results in a reduction in the discharge current.

The dependence of the discharge current on the microwave field level was also measured on the target probe (Fig. 39). As with these dependences for the current leaving the target (see Fig. 27), the current to the target saturates at powers above threshold. The measured current density at a power of P ≃ 0.2 MW and a target potential of 150 V was 0.1 A/cm^2. The current density did not change significantly at other powers.

It is interesting to observe the formation of a plasma at the target surface. For this purpose a potential of −300 V was applied to the collector of the target probe. An ion current was then observed to gradually increase during the microwave pulse. Figure 40 shows the relevant oscilloscope traces: above is the envelope of the microwave pulse and below is the ion current to the collector (U_0 = 100 V, P = 0.3 MW). Figure 41 shows the dependence of the amplitude of the ion current on the microwave power. The source of the plasma was evidently ionized gas driven out by electron bombardment of the target.

The reduction in the ion current density (see Fig. 41) with increasing target potential is natural since a positive charge on the target repels ions and the higher the potential is, the more strongly they are repelled.

17. An Analysis of the Factors Which Stabilize the Discharge and a Discussion of the Experimental Results

The main task of our investigations has been to determine the threshold conditions for development of a polyphase secondary-emission discharge. As we have been examining the initial stages of the development of

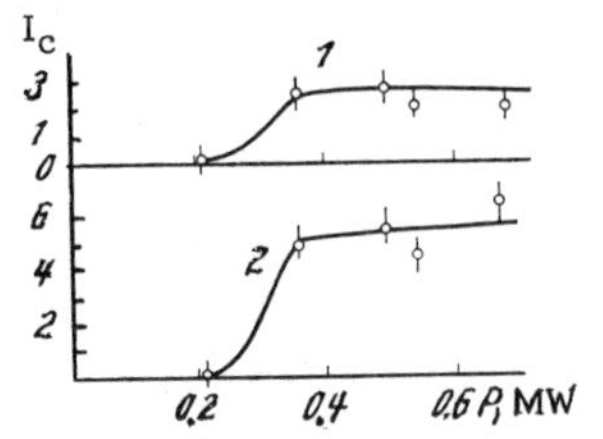

Fig. 39

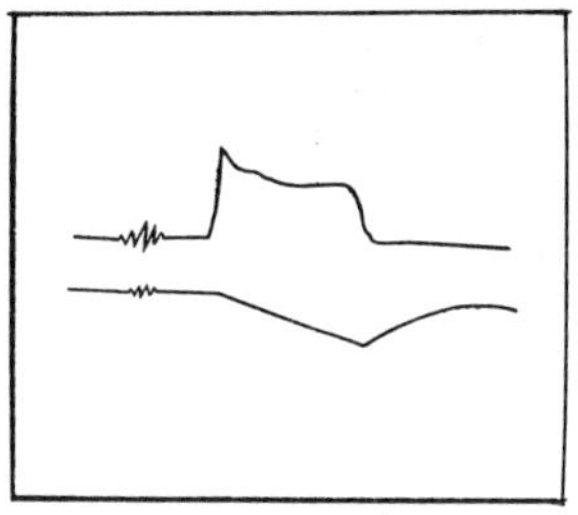

Fig. 40

Fig. 39. The current density to the target as a function of the microwave power for two target potentials: (1) $U_0 = 500$ V; (2) $U_0 = 300$ V.

Fig. 40. An oscilloscope trace of the ion current to the collector of the target probe. The upper trace is the envelope of the microwave pulse.

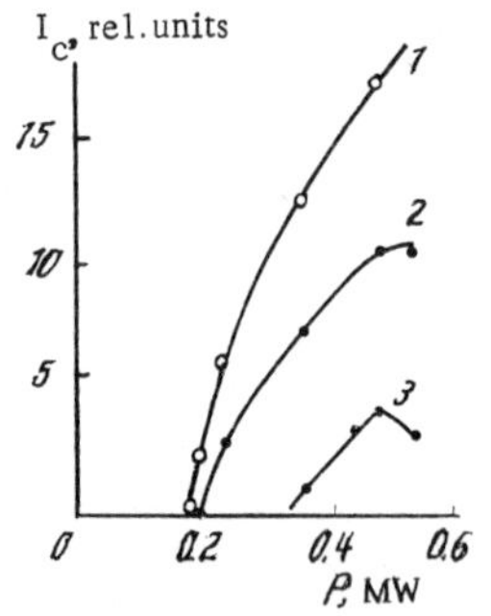

Fig. 41

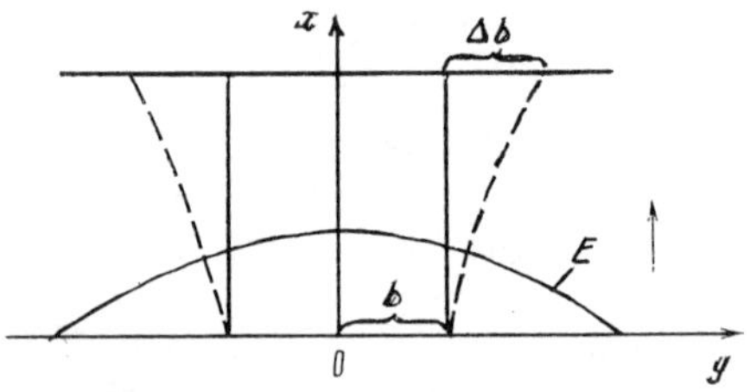

Fig. 42

Fig. 41. The ion current to the collector of the target probe as a function of the microwave power: (1) $U_0 = 60$ V; (2) = 150 V; (3) $U_0 = 440$ V.

Fig. 42. The geometry for which the factors tending to stabilize the discharge are evaluated.

the electron avalanche, we have neglected the effect of space charge, and the electron motion has been calculated without including the Coulomb interaction.

In this section we shall make a very approximate analysis of the role of the factors which stop the avalanche-type growth in the number of electrons participating in a polyphase discharge. We shall examine the dependence of these factors on the magnitude of the microwave field and on the magnitude of the restoring force in a one-sided discharge. The results of this analysis are compared with experimental data.

If the electron discharge occupies a limited region on the plane, it is possible to list the factors which stabilize the current density at a certain level. Obviously if the fraction of electrons that goes out of the discharge region under the action of the repulsive forces equals the relative growth in the number of electrons because $\langle\sigma\rangle > 1$ then the discharge current ceases to rise further. In our experiment (see Fig. 13) the factors which remove electrons from the discharge are the Coulomb field of the space charge and the Miller force.

We now evaluate the effect of these factors under the following simple geometric conditions (Fig. 42). Let a discharge occur between two parallel electrodes such that the alternating electric field perpendicular to the electrodes is distributed according to $E_m \cos(\pi y/a)$. Along the x and z axes the field is uniform. Let $E_m \times \cos(\pi b/a) = E_{thr}$, where E_{thr} is defined with the aid of Fig. 6.

We shall assume that electrons incident on the electrode with $|y| > b$ have no "descendants." Evidently a loss of electrons from the discharge zone is analogous to a reduction in the averaged emission coefficient $\langle\sigma\rangle_{inf}$ defined for infinite plane parallel electrodes, i.e.,

$$\langle\sigma\rangle' = \langle\sigma\rangle_{\text{inf}}\left(1 + \frac{n_1}{n_0}\frac{\Delta b}{b}\right)^{-1}, \tag{71}$$

where Δb is the deflection for an electron emitted at the boundary of the region b (see Fig. 42), n_0 is the mean electron density in the segment b, and n_1 is the mean electron density incident on the segment Δb.

It is easy to see that Eq. (71) is also valid for a one-sided discharge. In both two-sided and one-sided discharges the deviation Δb can be estimated as

$$\Delta b = \frac{F}{m}\frac{t_f^2}{2}, \tag{72}$$

where F is the force driving electrons from the discharge region and t_f is the time of flight of an electron.

In a one-sided discharge we can distinguish two cases:

$$U_0 > \frac{eE_m^2}{2m\omega^2}, \quad \text{when } t_f = 2\frac{E_m}{E_u}\frac{\sqrt{k(\beta,\gamma)}}{\omega} \tag{73}$$

[here $k(\beta, \gamma) = W_{avg}/W$ (see Fig. 7)] and

$$U_0 < \frac{eE_m^2}{2m\omega^2}, \quad \text{when } t_f = 2d\sqrt{\frac{2m}{eU_0}} \tag{74}$$

(here d is the characteristic distance over which the constant field of the target E_u extends and $E_u d \simeq U_0$; when the target is in the center of the waveguide, d is the distance to the waveguide wall). It remained to determine the quantity F in Eq. (72) which is the sum of the Miller force and the electrostatic repulsive space-charge force.

It is easy to show that if $b \ll a$, then at a distance b from the axis the Miller force will be

$$F_M = \frac{e^2E_m^2}{2m\omega^2}\left(\frac{\pi}{a}\right)^2 b. \tag{75}$$

For simplicity we estimate the Coulomb repulsion as

$$F_C = e^2 n_0 b/\varepsilon_0 \tag{76}$$

(here ε_0 is the vacuum dielectric constant). We assume that the force F_C is produced by the electric field of a charged plane of thickness 2b. Substituting the sum of Eqs. (75) and (76) into Eq. (72) together with the value of t_f from Eqs. (73) and (74), from Eqs. (71) and (72) we obtain

$$\langle\sigma\rangle' = \langle\sigma\rangle_{\text{inf}}\left\{1 + \frac{n_1}{n_0}\left[k\left(\pi\frac{r}{a}\frac{E_m}{E_u}\right)^2 + 2\frac{E_m}{E_u^2}\frac{I\sqrt{k}}{\varepsilon_0\omega}\right]\right\}^{-1} \tag{77}$$

if $U_0 > eE_m^2/2m\omega^2$, where $r = eE_m/m\omega^2$ and $I = en\sqrt{2kW/m}$ is the current density; if $U_0 < eE_m^2/2m\omega^2$, then

$$\langle\sigma\rangle' = \langle\sigma\rangle_{\text{inf}}\left\{1 + \frac{n_1}{n_0}\left[\frac{W}{eU_0}\left(2\pi\frac{d}{a}\right)^2 + 4\frac{I}{\varepsilon_0}\sqrt{\frac{dm}{2E_u^3 e}}\right]\right\}^{-1}. \tag{78}$$

Equations (77) and (78) make it possible to qualitatively (since n_1/n_0 is not known) estimate the extent to which the perpendicular forces reduce the average electron multiplication coefficient. Evidently the growth in the discharge current is stabilized when $\langle\sigma\rangle' = 1$.

The potential fields can change $\langle\sigma\rangle'$ in two ways: either by inhibiting (easing) the escape of electrons from the discharge region or by changing the discharge conditions at the emitting surface. The first possibility was examined previously [see Eqs. (77) and (78)]. We now consider the second. The effect of the restoring force f on the nature of the discharge is obvious since by changing $f = eE_u$ it is possible to realize any of the three types of secondary-emission discharge (see Sec. 1).

We now treat the simplest situation in which an alternating electric field that is greater than the threshold field is perpendicular to an infinite plane. In this case as the electrostatic field at the target surface E_M is gradually increased a field E_{cr_1} will be reached at which the conditions for a uniform polyphase SED occur. As E_M is increased further it becomes kinetically possible for a polyphase SED to exist (true, it is unclear whether the available alternating electric field is sufficient to maintain the discharge under these conditions), but with a further increase in E_M a situation surely arises in which a discharge is impossible. This happens when the discharge is substantially of a resonant character but the kinematic conditions for existence of an

SERD are not satisfied at the given E_M. We denote the minimum value of E_M at which a discharge cannot exist by E_{cr_2}.

When a discharge exists the field at the surface of the target is made up of the two components, E_u and E_q, of the space-charge field, i.e.,

$$E_M = E_u + E_q.$$

If $E_u \gg E_q$ then the distance by which the electrons are deflected from the surface is $l \simeq W/eE_u$. Then

$$E_M = E_u + \frac{n_e W}{2\varepsilon_0 E_u}. \tag{79}$$

Clearly the density n_e will increase until $\langle\sigma\rangle'$ falls to 1, which happens when $E_M = E_{cr_2}$.

Using the relation $n_e = (I/e)\sqrt{m/W}$, we can obtain from Eq. (79) an expression relating the current density at the target, I, to the constant field at the target E_u and to the energy of an electron in the wave field:

$$I = \frac{4\varepsilon_0 e E_u (E_{cr_2} - E_u)}{\sqrt{2Wm}}. \tag{80}$$

This formula is valid only when $E_{cr_2} - E_u \ll E_{cr_2}$. In the other limiting case where the field at the target is determined mainly by the space charge $E_{cr_2} - E_u \sim E_{cr_2}$ and it is easy to obtain

$$I = \frac{4\varepsilon_0}{d}(E_{cr_2} - E_u)\sqrt{\frac{2W}{m}}. \tag{81}$$

Equation (81) was derived under the assumption that on the average an electron is acted on by half the field strength found at the target surface.

It is clear from Eqs. (80) and (81) that when E_u is increased while the microwave field strength is kept constant we should expect a drop in the discharge current density I. Equation (81), which gives only a qualitative description of the dependence, is in fully satisfactory agreement with the experimental results [see Fig. 38 where the straight line is calculated according to Eq. (81)]. When $U_0 \sim U_{thr_2}$ Eq. (81) is not applicable and Eq. (80) must be used in this range.

Clearly the discharge current density can be estimated using Eq. (81). Thus, for P = 363 kW and $E_{cr_2} \approx$ 70 kV/m, we find I = 0.6 kA/m^2. For this case it was found experimentally (see curve 1 of Fig. 39) that I = 1 kA/m^2, which may be regarded as satisfactory agreement in view of the very approximate nature of Eq. (81).

It would be interesting to determine how the current density to the target depends on the microwave power level. To do this we could use Eqs. (77), (78), (80), and (81). However, we do not know which of the discharge stabilizing factors – the transverse scatter or the change in the nature of the discharge due to the space-charge field – is stronger. According to Eqs. (77), (78), and (80) the current should fall as the microwave power is increased, while according to Eq. (81) it should increase in proportion to $\sqrt{W}$. Since in addition to the factors accounted for in Eq. (81) there are factors which act to reduce the current as the power is increased, the result of our qualitative analysis may be stated as follows: as the microwave field is increased the discharge current will not rise faster than the square root of the power. This conclusion is in agreement with experiment (see Fig. 39).

CONCLUSIONS

We now summarize the main results of this paper.

1. We have analyzed the effect of the actual scatter in the initial velocities of the secondary-emission electrons. It has been shown that, as a consequence of this factor, it is possible to identify three types of secondary-emission microwave discharges in terms of the nature of the distribution of in-phase electrons in front of the discharge surface: monophase, polyphase, and uniform polyphase. The ranges in which these types exist were pointed out. It is shown that the existing theory of the SERD – the resonant discharge – is valid only for the first type of discharge.

2. The third type of discharge, the uniform polyphase SED, is typical for systems with a larger time interval between the moment of emission and a new collision compared to the period of the field. The problem has been solved analytically for this type of discharge under the assumption that σ is directly proportional to the energy. The distribution function of the electrons has been found and demonstrated to be stable. The threshold criterion for a discharge has been found and the growth rate determined. This method has been also used

to determine the steady-state distribution function and the criterion for a discharge when there is an arbitrary angle between the electric field vector of the microwaves and the surface.

3. It has been shown that the entire set of possible distribution functions for arbitrary values of the alternating electric field is completely determined by just two parameters β and γ if the secondary-emission characteristic of the material can be approximated by two straight lines. A computer calculation was used to find the distribution functions of the discharge current with respect to the phase of emission and to find the conditions for development of a uniform polyphase SED.

These results are presented in the form of graphs which can be used for given β and γ to find the appropriate distribution function and range of microwave fields within which a self-sustaining uniform polyphase discharge can exist on a given substance.

4. We have determined experimentally the threshold microwave fields at which a polyphase SED can develop for a number of materials with different secondary-emission characteristics. Good agreement was found between the experimental and calculated threshold microwave fields.

5. A qualitative study has been made of the relationship between the magnitude of the restoring force required for development of a self-sustaining uniform polyphase SED and the secondary emission properties of a material when the electric field of the microwaves is perpendicular to the surface and the secondary emission coefficient is directly proportional to the energy.

It is shown that the threshold restoring force is proportional to the first critical potential and depends only weakly on the electric field of the microwaves.

We have studied the relationship between the secondary-emission characteristics, the magnitude of the electric component of the microwave field, and the magnitude of the restoring force required for a discharge to occur. Experimental measurements were made of the current density in a uniform polyphase secondary-emission discharge as a function of the restoring force and of the rf electric field.

Satisfactory agreement was obtained between the theoretical calculations and the measurements.

As a result of this work we have established the basic features of the initial stage of a discharge of the sort characteristic of systems with a large time of flight for the electrons.

The dependence of the current density in the developed stage of a discharge on two principal factors, the restoring force and the microwave field, has been determined experimentally.

APPENDIX

DETERMINATION OF THE OPERATOR FOR FINDING THE DISTRIBUTION OF THE SECONDARY CURRENT FROM A GIVEN PRIMARY CURRENT DISTRIBUTION FUNCTION

In Sec. 2 it was shown that if there is a uniform distribution of the density of particles emitted in phase φ_1 in the space in front of a wall and there is a time average current I_1 of these particles, then the result is a current to the wall that varies with the phase φ as

$$i_1(\varphi, \varphi_1) = \frac{\cos\varphi + \cos\varphi_1}{\cos\varphi_1} I_1(\varphi_1). \tag{A.1}$$

Depending on the phase of emission φ_1, the range of phases of the field where an electron can collide with the wall encompasses a larger or smaller range of phases φ. The limits of this range are shown in Fig. 2. Electrons emitted with phases of 0 to $\pi/2$ can leave the wall. Only these electrons can participate in the discharge; consequently, we are interested in the distribution with respect to the phase of arrival of primary electrons only over the phase interval from 0 to $\pi/2$. In Sec. 2 it was shown that electrons emitted within the phase interval 0-1.35 yield a cosinusoidal distribution of the type (A.1) over the entire active interval 0-1.57 when they arrive at the wall. Those electrons that are emitted over the range 1.35-1.57 and move toward the wall will strike it in accordance with Eq. (A.1) only when $\varphi > \alpha(\varphi_1)$ (where $0 < \alpha < \pi/2$).

The phase interval $0 - \alpha(\varphi_1)$ that is free of incident electrons with emission phases $1.35 < \varphi_1 < 1.57$ will be larger, the closer φ_1 is to 1.57. In the limit $\alpha = \pi/2$ for $\varphi_1 = 1.57$.

Now that we have obtained an idea of the phase distribution of the field at the time the primary electrons arrive, we consider the distribution of the secondaries. We shall assume that secondary electrons are emitted instantaneously after impact of the primaries.

If the energy of an incident electron $W(\cos\varphi + \cos\varphi_1)^2$ is less than the energy W_b, then because σ is directly proportional to the impact energy the secondary current induced by the current $I_1(\varphi_1)$ will be distributed over the phase of emission φ according to Eq. (A.1) as follows:

$$i(\varphi, \varphi_1) = \sigma_b \frac{W}{W_b}(\cos\varphi + \cos\varphi_1)^2 \frac{\cos\varphi + \cos\varphi_1}{\cos\varphi_1} I_1(\varphi_1). \tag{A.2}$$

If, however, the energy of the incident electrons is greater than W_b, then (as can easily be seen from Fig. 4)

$$i(\varphi, \varphi_1) = \sigma_b \frac{1-(W/W_0)(\cos\varphi + \cos\varphi_1)^2}{1 - W_b/W_0} \frac{\cos\varphi + \cos\varphi_1}{\cos\varphi_1} I_1(\varphi_1). \tag{A.3}$$

The secondary current emitted in any phase $0 < \varphi < \pi/2$ is made up of the secondary currents $i(\varphi, \varphi_1)$ produced by primary currents $i_1(\varphi_1)$ with emission phases of $0 < \varphi_1 < 1.35$ and $1.35 < \varphi_1 < 1.57$, although the second group of phases of emission does not contribute over all phases φ.

Let us break up the interval 0–$\pi/2$ into n segments. We introduce the coefficient k_{lp} which relates the primary current in phase interval p to that portion of the secondary current in interval l caused by it. Since the time-averaged current $I_1(\varphi_1) = i_1(\varphi_1)/4n$ [where $i_1(\varphi_1)$ is the primary current in the middle of interval p], Eqs. (A.2) and (A.3) imply that the coefficient k_{lp} will have the form

$$k_{lp} = \sigma_b \frac{W}{W_b}\left[\cos\frac{\pi(l-1/2)}{2n} + \cos\frac{\pi(p-1/2)}{2n}\right]^3 \frac{1}{4n}\left[\cos\frac{\pi(p-1/2)}{2n}\right]^{-1}, \tag{A.4}$$

if $\left[\cos\frac{\pi(l-1/2)}{2n} + \cos\frac{\pi(p-1/2)}{2n}\right]^2 < \beta$;

$$k_{lp} = \sigma_b\left[1 - \frac{\gamma}{\beta}\left(\cos\frac{\pi(l-1/2)}{2n} + \cos\frac{\pi(p-1/2)}{2n}\right)^2\right]\left[\cos\frac{\pi(l-1/2)}{2n} + \cos\frac{\pi(p-1/2)}{2n}\right]\left[4n(1-\gamma)\cos\frac{\pi(p-1/2)}{2n}\right]^{-1}, \tag{A.5}$$

if $\left[\cos\frac{\pi(l-1/2)}{2n} + \cos\frac{\pi(p-1/2)}{2n}\right]^2 > \beta$; and

$$k_{lp} = 0,$$

if $\varphi < \alpha(\varphi_1)$.

If the distribution of the primary current with respect to the phase of emission is known, then the distribution of the secondary current with respect to the phase of emission can be found using an $n \times n$ matrix whose elements are the coefficients k_{lp} (l is the column number; p the row number).

In order to do this, the primary current distribution function must be written in the form of a set of n discrete values equal to the values of the primary function in the middle of each (p-th) segment $i_1(p)$ over the interval 0–$\pi/2$. Then the p-th element of the function is multiplied by each element of the p-th row of the matrix. Adding the resulting values in each column l, we obtain the distribution function of the secondary current in each cell l of the interval 0–$\pi/2$. Later on, using this matrix and knowing the distribution function of the secondary current, one can find the distribution function of the tertiary current, etc.

In this paper, n was equal to 10. A stable function with an accuracy of 10^{-3} was obtained after 5-10 cycles. The calculation was done on a "Mir-1" computer with an arbitrarily chosen primary function.

LITERATURE CITED

1. I. R. Gekker and O. V. Sizukhin, Pis'ma Zh. Éksp. Teor. Fiz., 9:403 (1969).
2. K. F. Sergeichev and V. E. Trofimov, Pis'ma Zh. Éksp. Teor. Fiz., 13:236 (1971).
3. G. M. Batanov and K. A. Sarksyan, Tr. Fiz. Inst. Akad. Nauk SSSR, 73:104 (1973).
4. W. Henneburg, R. Orthuber, and E. Stendel, Z. Tech. Phys., 17:115 (1936).
5. A. J. Hatch, J. Appl. Phys., 32:1086 (1961).
6. V. G. Mudrolyubov and V. I. Petrunin, Zh. Tekh. Fiz., 10:1017 (1970).
7. I. M. Bronshtein and B. S. Freiman, Secondary Electron Emission [in Russian], Nauka, Moscow (1969).
8. A. Miller, H. Williams, and O. Theimer, J. Appl. Phys., 34:1674 (1963).

9. A. Miller, Dissertation, New Mexico State University (1961).
10. B. A. Zager and V. G. Tishin, Zh. Tekh. Fiz., 33:1121 (1963).
11. B. A. Zager, Prib. Tekh. Éksp., No. 2, 20 (1963).
12. I. A. Kossy, G. S. Lukianchikov, and M. M. Savchenko, XI Internat. Conf. on Phenomena in Ionized Gases, Contributed Papers, Prague (1973), p. 93.
13. V. I. Alexandrov, D. A. Ganichev, and V. I. Presnov, XI Internat. Conf. on Phenomena in Ionized Gases, Contributed Papers, Prague (1973), p. 94.
14. V. S. Gorovets and S. A. Rybak, Zh. Tekh. Fiz., 37:387 (1967).
15. V. A. Stanskii, D. A. Ganichev, and S. A. Fridrikhov, Zh. Tekh. Fiz., 43:1750 (1973).
16. V. P. Sazonov, B. V. Prokof'ev, and G. M. Priezzhaev, Élektron. Tekh., Ser. 1, No. 12, 3 (1970).
17. L. N. Dobretsov, Electron and Ion Emission [in Russian], Gostekhizdat, Moscow (1952).
18. S. Ramo and G. Winnery, Fields and Waves in Modern Electronics [Russian translation], Gostekhizdat, Moscow (1950).
19. G. S. Luk'yanchikov, Zh. Tekh. Fiz., 44:1922 (1974).
20. L. V. Grishin and G. S. Luk'yanchikov, Zh. Tekh. Fiz., 46:536 (1976).
21. A. A. Dorofeyuk, I. A. Kossyi, G. S. Luk'yanchikov, and M. M. Savchenko, Zh. Tekh. Fiz., 46:138 (1976).